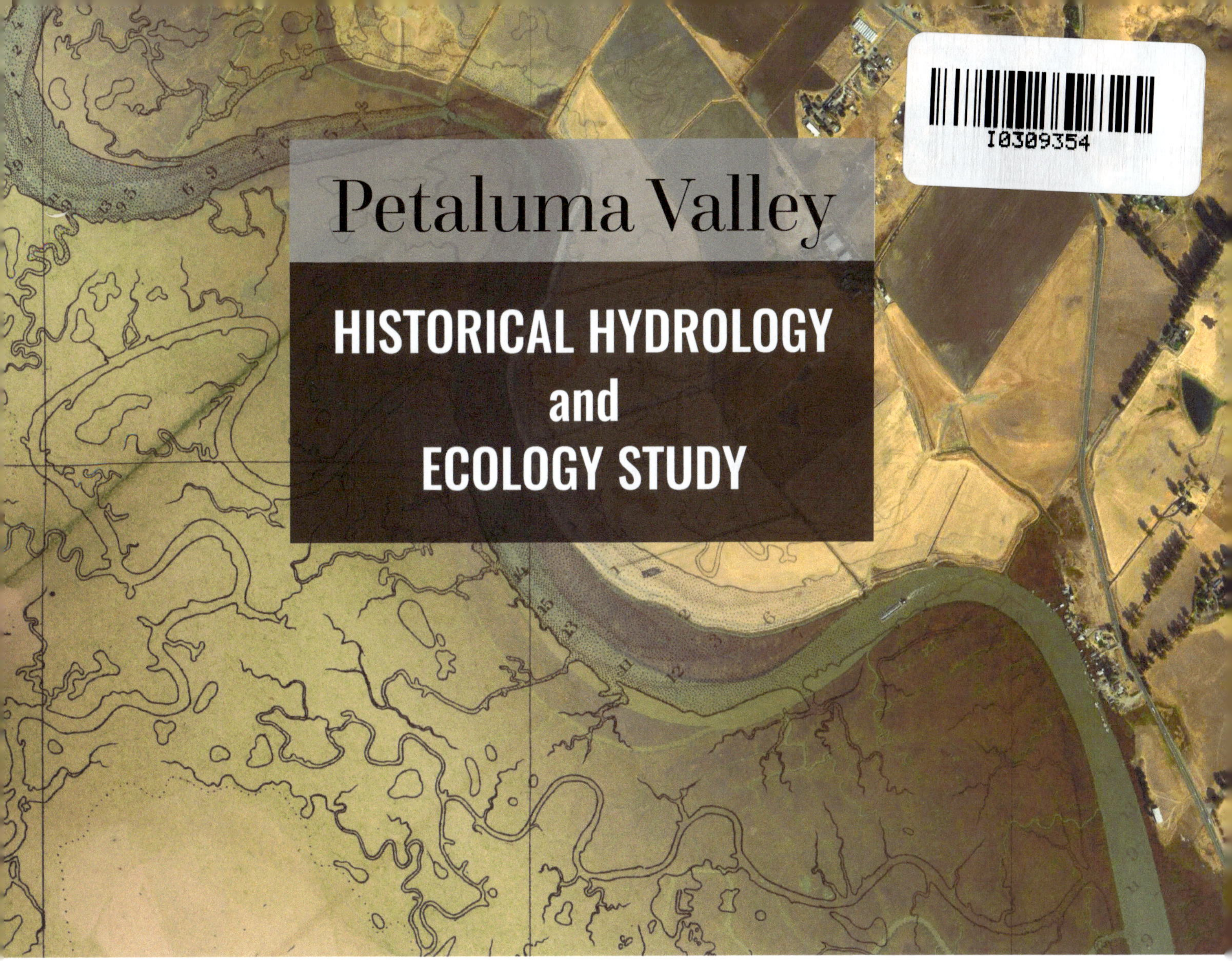

Petaluma Valley

HISTORICAL HYDROLOGY and ECOLOGY STUDY

PREPARED BY

Sean Baumgarten
Emily Clark
Scott Dusterhoff
Robin Grossinger
Ruth Askevold

San Francisco Estuary Institute-Aquatic Science Center

SAN FRANCISCO ESTUARY INSTITUTE PUBLICATION #861

March 2018

SUGGESTED CITATION

Baumgarten SA, Clark EE, Dusterhoff SR, Grossinger RM, Askevold RA. 2018. Petaluma Valley Historical Hydrology and Ecology Study. Prepared for the Sonoma Resource Conservation District and U.S. Environmental Protection Agency. A Report of SFEI-ASC's Resilient Landscapes Program, SFEI Publication #861, San Francisco Estuary Institute, Richmond, CA.

VERSION

March 2018 (v1.0)

REPORT AVAILABILITY

Report is available on SFEI's website at www.sfei.org/projects/petaluma-valley-historical-hydrology-and-ecology-study

IMAGE PERMISSION

Permissions for images used in this publication have been specifically acquired for one-time use in this publication only. Further use or reproduction is prohibited without express written permission from the responsible source institution. For permissions and reproductions inquiries, please contact the responsible source institution directly.

FUNDING

This project has been funded wholly or in part by the United States Environmental Protection Agency under assistance agreement 99T34001 to the Sonoma Resource Conservation District. The contents of this document do not necessarily reflect the views and policies of the Environmental Protection Agency, nor does the EPA endorse trade names or recommend the use of commercial products mentioned in this document.

COVER CREDITS

Historical map from 1861 overlaid on contemporary imagery. (USCS 1861, courtesy of NOAA; NAIP 2016, courtesy of USDA)

CONTENTS

EXECUTIVE SUMMARY — 1
- Historical Landscape — 2
- Landscape Change — 2

1. INTRODUCTION — 5
- Project Background — 5
- Report Structure — 7
- Why Historical Ecology? — 7
- Environmental Setting — 8
- Land Use Context — 10

2. METHODOLOGY — 15
- Data Collection and Compilation — 15
- GIS Synthesis and Mapping — 19
- Analysis of Change over Time — 20
- Identification of Potential Restoration Opportunity Areas — 21
- Transferability of Methods — 21
- Technical Advisory Committee — 22

3. ESTUARY — 23
- Overview — 23
- Petaluma River and Tidal Channels — 26
- Marsh Plain — 30
- Transition Zone — 32
- Fish and Wildlife — 34

4. STREAMS AND RIPARIAN HABITATS — 37
- Overview — 37
- Flooding and Flow Variability — 39
- Hydrologic Connectivity — 43
- Riparian and Aquatic Habitats — 47

5. NON-TIDAL WETLANDS — 55
- Overview — 55
- Wet Meadow — 56
- Vernal Pool Complex — 60
- Valley Freshwater Marsh — 64
- Willow Grove — 67

6. CHANGE OVER TIME	**69**
Overview	69
Tidal Marsh Diked, Drained, and Filled	73
Petaluma River Dredged and Straightened	76
Loss of Seasonal Wetlands	80
Draining of Laguna de San Antonio	82
Channelization and Lengthening of Streams	85
Changes in Channel Alignment	87
7. SYNTHESIS AND NEXT STEPS	**89**
Feasibility Analysis	93
Watershed Restoration Vision	93
REFERENCES	**94**
APPENDIX: SPECIES NAMES	**105**

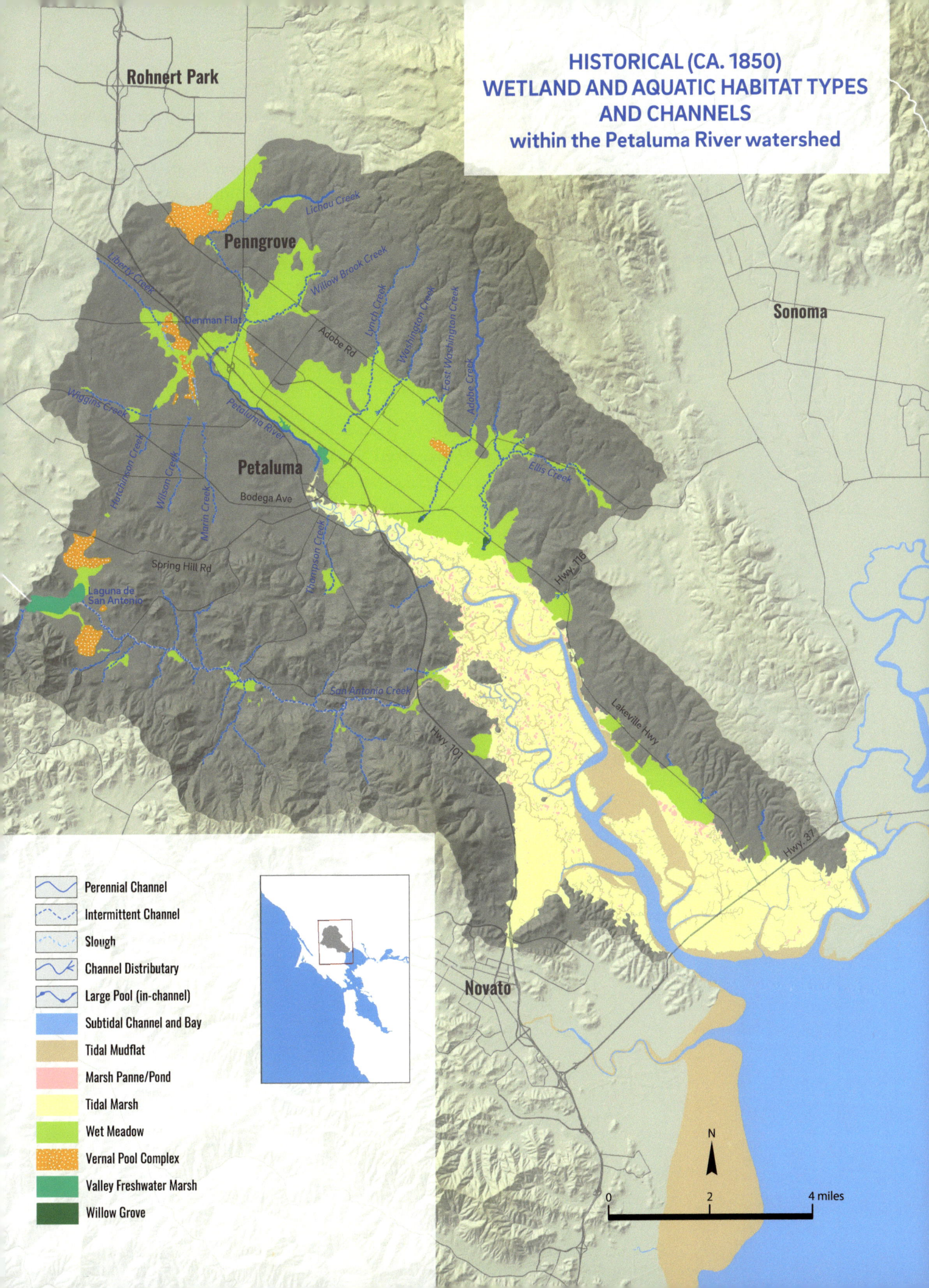

ACKNOWLEDGEMENTS

This project was funded by the U.S. Environmental Protection Agency (EPA) and was conducted in partnership with the Sonoma Resource Conservation District (SRCD). We would like to thank Jared Vollmer (EPA) and Anya Starovoytov (SRCD) for their support and guidance throughout the course of the project. Valerie Minton and Kara Heckert (SRCD) and Chuck Striplen (North Coast Regional Water Quality Control Board, formerly SFEI-ASC) envisioned the project and worked to see it come to fruition.

Our Technical Advisory Committee (TAC) helped to ensure the technical accuracy of the research by providing many helpful comments and suggestions on methodology, emerging findings, and the draft report. Thank you to TAC members Betty Andrews, Stephanie Bastianon, Jason Beatty, Laurel Collins, John Fitzgerald, Susan Haydon, Judy Kelly, John McKeon, Julian Meisler, Andy Rodgers, Solange Russek, Nancy Scolari, Gail Seymour, William Stockard, Chase Takajo, Jared Vollmer, Jennifer Walser, and Jason White. Laurel Collins (Watershed Sciences) contributed additional advice and guidance to our analysis of potential restoration opportunity areas.

We are indebted to all of the staff and volunteers at invaluable archives visited during the course of the project. In particular, we would to thank Brian and Bob Curtis at Curtis & Associates, Inc.; Solange Russek and John Fitzgerald at the Petaluma Historical Library & Museum; Dewey Livingston at the Anne T. Kent California Room at the Marin County Free Library; Pat Keats and Heather Fordham at the Society of California Pioneers; Katherine Rinehart at the Sonoma County History and Genealogy Library; Matt Fossum at the California State Lands Commission; Gina Bardi at the San Francisco Maritime National Historical Park Research Center; Nancy Wilson at the Petaluma History Room at the Sonoma County Library; and Lynn Prime at the Sonoma State University Library. Thank you as well to the landowners who generously allowed us to conduct site visits during the course of the project.

Numerous SFEI staff assisted with data collection, compilation, mapping, analysis, reporting, and project management, including Erin Beller, Josh Collins, Letitia Grenier, Steven Hagerty, Pete Kauhanen, Jeremy Lowe, Sarah Lowe, Amy Richey, and Micha Salomon. Two interns from the Bill Lane Center for the American West at Stanford University—Kate Roberts and Miranda Vogt—also provided valuable research assistance.

EXECUTIVE SUMMARY

Petaluma River. *(Photo by Carolyn Jewel, December 2017, licensed under Creative Commons)*

This study examines the historical hydrology and ecology of the Petaluma River watershed prior to major Euro-American modification, and analyzes landscape changes over the past two centuries. Synthesizing information from hundreds of archival documents, the research reconstructs the historical form and function of wetland, riparian, and aquatic habitats and stream channels throughout the watershed, providing insights into habitat extent and distribution, streamflow and sediment dynamics, vegetation composition, wildlife support, and landscape change. Findings from this research can be used to help set restoration targets and to prioritize multi-benefit opportunities to restore wildlife habitat, enhance flood protection, increase groundwater recharge, and improve sediment management. This Executive Summary highlights key findings from the study.

HISTORICAL LANDSCAPE

Tidal wetlands occupied 6,540 ha (16,150 ac) along the lower Petaluma River. The tidal wetlands were composed of a range of estuarine habitat types including tidal marsh, tidal mudflats, subtidal channels, and marsh ponds/pannes. The Petaluma River entered the estuary near present-day Payran Street, and followed a sinuous course for 28 km (17 mi) to its mouth at San Pablo Bay. Influenced both by tidal flux and by freshwater input from the Petaluma River, San Antonio Creek, and other tributaries, the tidal wetlands formed a dynamic landscape that supported a wide variety of plants and animals, including presently threatened or endangered species such as Ridgway's rail, salt marsh harvest mouse, and soft bird's beak. Tidal-terrestrial transition zones bordered the estuary, forming a link between the tidal wetlands and the adjacent upland and fluvial habitats.

Non-tidal wetlands occupied 4,610 ha (11,400 ac) in valley floor settings throughout the watershed. Seasonal wetlands such as wet meadow and vernal pool complex formed in areas that received seasonal freshwater inputs from rainfall, flooding, or groundwater, while perennial wetlands such as valley freshwater marsh and willow grove occupied areas with perennial standing water or saturated soils. The largest contiguous non-tidal wetland feature was a wet meadow that occupied much of the alluvial plain on the east side of the Petaluma River. Large wetland complexes existed at the head of San Antonio Creek (the Laguna de San Antonio) and in the Denman Flat area near the head of the Petaluma River. Non-tidal wetlands provided important habitat for amphibians, migratory waterfowl, currently imperiled species such as tricolored blackbird, and many other wildlife.

Upstream of the estuary, the Petaluma River was characterized by a short, relatively straight single-threaded channel with large pools. The river exhibited a high degree of seasonal flow variability: flows were minimal during the dry season, but during the wet season floods periodically inundated large areas along the mainstem and on the alluvial plain to the east. A number of large in-channel pools maintained perennial water and may have provided cold-water refugia for salmonids and other native fish during the dry season.

Streamflow patterns and degree of hydrologic connectivity in tributary channels varied both spatially and temporally. Streamflow in the lower portions of many tributaries was intermittent, though in some cases upstream springs and wetland complexes maintained limited perennial reaches. Smaller tributaries with lower stream power generally did not maintain defined channels to the estuary or the Petaluma River. Notable exceptions included Lichau Creek, which connected directly to the Petaluma River mainstem, and San Antonio Creek, which flowed directly into the tidal San Antonio Slough. Many stream channels supported mixed riparian forests dominated by oaks, willows, alders, and California bay laurel.

LANDSCAPE CHANGE

The tidal portion of the Petaluma River was dredged and straightened in order to make the river more conducive to maritime navigation. Channel modifications commenced in the late 19th century with large-scale dredging and the construction of numerous cut-offs, and sediment accumulation within the channel necessitated ongoing dredging throughout the 20th century. Sediment accretion has also raised the elevation of hundreds of acres of former tidal channels and mudflats at the mouth of the Petaluma River and at False Bay, much of which converted to tidal marsh (and then was often diked).

The area of tidal wetland types has decreased by 58%. Beginning in the late 19th century, thousands of acres of tidal marsh were diked and drained in an effort to reclaim lands for agricultural use. Construction of transportation corridors and industrial infrastructure further contributed to tidal wetland loss. Despite the substantial loss of tidal wetland habitats, the Petaluma Marsh remains the largest contiguous expanse of historical tidal marsh in San Pablo Bay. Restoration efforts in recent decades have begun to reverse the decline in tidal wetland extent.

The area of non-tidal wetland types has decreased by 84%. The large wet meadow that occupied much of the valley floor east of the Petaluma River has been almost completely eliminated, as have the vast majority of vernal pool complexes throughout the watershed. Though the Laguna de San Antonio wetland complex was ditched and drained in the late 19th century, modified and remnant wetlands still occupy approximately 55 ha (140 ac) at the head of San Antonio Creek.

Many stream segments have been channelized and straightened to increase drainage efficiency and control flooding. Streams that were historically disconnected from the estuary or mainstem channel downstream have been lengthened, and today artificial channels convey flows and sediment further downstream than they did in the past. Portions of many tributaries—most notably lower San Antonio Creek—have also been realigned. In addition, thousands of feet of artificial channels have been constructed through diked baylands to facilitate drainage. These modifications have resulted in a 50% increase in channel length among higher order channels within alluvial areas.

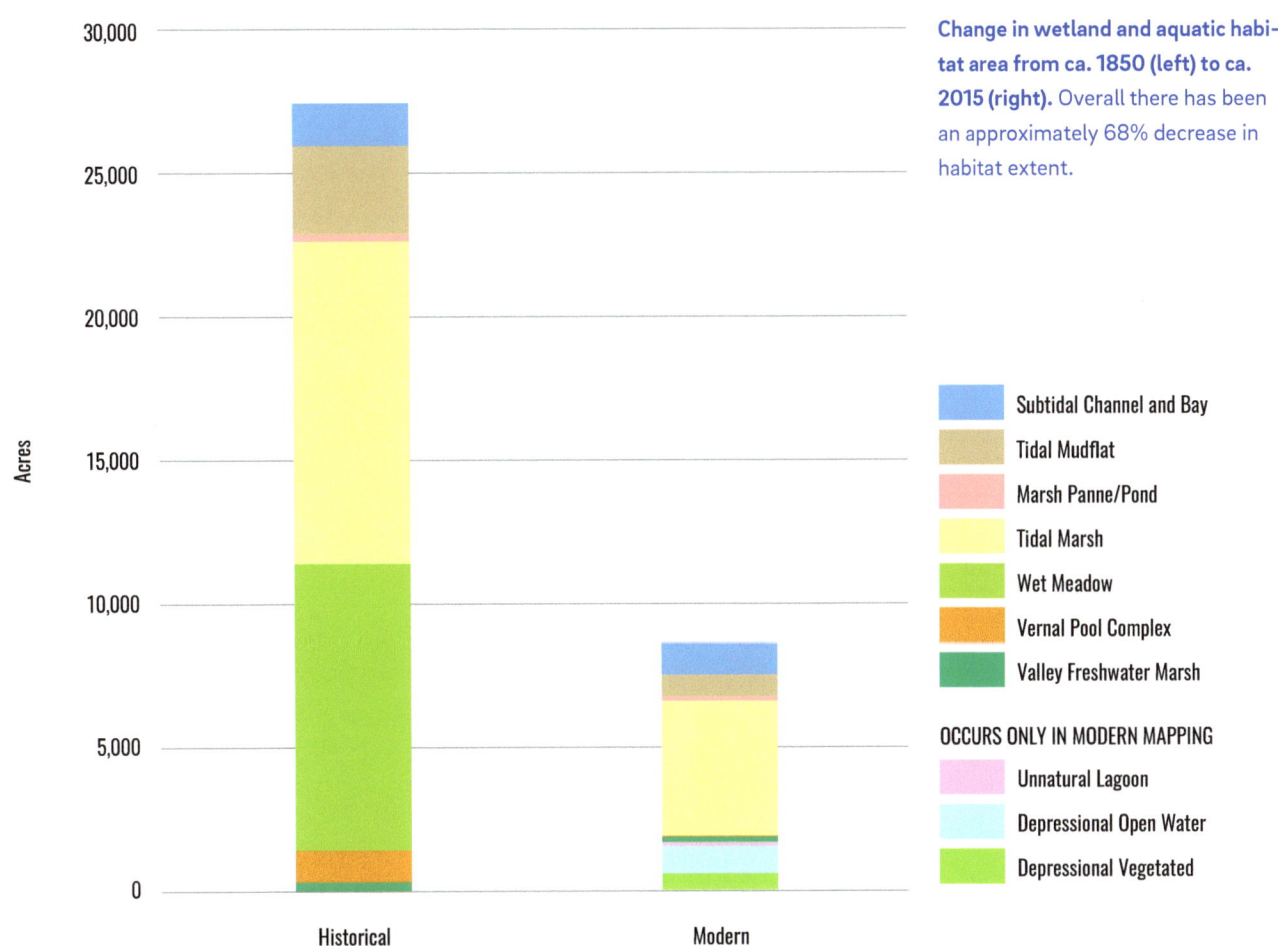

Change in wetland and aquatic habitat area from ca. 1850 (left) to ca. 2015 (right). Overall there has been an approximately 68% decrease in habitat extent.

The steamer "Gold" on the Petaluma River (undated). *(Photo number B7.9,465pl, courtesy of San Francisco Maritime National Historical Park Research Center)*

1. INTRODUCTION

PROJECT BACKGROUND

This study reconstructs the historical landscape of the Petaluma River watershed and documents the major landscape changes that have taken place within the watershed over the past two centuries. Prior to Spanish and American settlement of the region, the Petaluma River watershed supported a dynamic and interconnected network of streams, riparian forests, freshwater wetlands, and tidal marshes. These habitats were utilized by a wide range of plant and animal species, including a number of species that are today listed as threatened or endangered such as Ridgway's Rail, Black Rail, salt marsh harvest mouse, California red-legged frog, Central California Coast steelhead, and soft bird's beak (CNDDB 2012, SRCD 2015). Agricultural and urban development beginning in the mid-1800s has significantly altered the landscape, degrading habitat for fish and wildlife and contributing to contemporary management challenges such as flooding, pollutant loading, erosion, and sedimentation. While many natural areas and remnant wetlands still exist throughout the watershed—most notably the Petaluma Marsh—their ecological function is in many cases seriously impaired and their long-term fate jeopardized by climate change and other stressors. Multi-benefit wetland restoration strategies, guided by a thorough understanding of landscape history, can simultaneously address a range of chronic management issues while improving the ecological health of the watershed, making it a better place to live for both people and wildlife.

A wide range of factors threaten both human and natural communities in the watershed over the coming decades, and pose serious challenges for land managers. Flood hazards exist in a number of developed areas of the watershed, such as along the Petaluma River mainstem upstream of the D Street Bridge, within the Denman Flat area near the head of the Petaluma River, and within the lower reaches of Willow Brook Creek. A number of recent and ongoing planning efforts seek to mitigate the impacts of flooding in flood-prone areas of the watershed; these include projects that integrate flood protection with other management objectives such as groundwater recharge and habitat restoration (SCWA 1986, Kennedy et al. 2012, City of Petaluma 2015, SRCD 2015). Erosion and sedimentation, driven by a combination of urban and agricultural development, vegetation removal, and hydrologic changes, have been identified as management priorities along many creeks throughout the watershed, including Willow Brook, Lynch, Adobe, Ellis, and San Antonio creeks (SSCRCD 2008, SRCD 2015). Water quality within the Petaluma River is also impaired: the river is currently listed under Clean Water Act Section 303(d) for bacteria, nutrients, diazinon, trash, and sediment (SFBRWQCB 2016, Ghodrati and Lunde 2017).

Sea level rise, projected to reach 0.4–1 m (1.6–3.4 ft), and possibly as much as 3 m (10 ft), in San Francisco Bay by 2100 under existing emissions trajectories (Griggs et al. 2017), puts existing wetlands in the Petaluma Marsh at risk and jeopardizes land uses in low-lying surrounding areas. Levee overtopping will likely become more frequent in diked, subsided baylands, while the increased frequency of inundation may accelerate bank erosion and habitat conversion in tidal wetlands (Goals Project 2015). Climate change will also alter streamflow patterns and vegetation distribution throughout the watershed; countywide, climate change is projected to increase the severity of flood events, the frequency and severity of droughts, and the frequency of extreme heat events (Cornwall et al. 2014).

Despite two centuries of diking and filling in many parts of estuary, Petaluma Marsh remains the largest and least disturbed remnant of the vast areas of brackish and saline tidal marsh that historically existed in the San Francisco Estuary. As such, Petaluma Marsh is broadly recognized as a primary source of information about the nature of mature tidal marsh ecosystems, and serves as the best existing natural analogue for what large-scale tidal marsh restoration efforts might achieve and how the resulting marshlands might be managed for their abundant benefits. Studies of its hydrology and ecology have been foundational to designs and plans for tidal marsh restoration in the San Francisco Estuary and beyond. The planform of the marsh, with its dendritic channel networks, natural levees, and tidal marsh pannes, has been used to calibrate the most detailed historical Topographic Sheets of the first Coast Survey (Grossinger et al. 2005), and to help explain the evolution of tidal marsh habitats (Collins et al. 1986 and 1987, Leopold et al. 1993, Collins and Grossinger 2004). Its mostly undisturbed marsh plains contain a record of sedimentation spanning more than two millennia that has been used to document long-term rates of sea level rise and its ecological effects (Byrne et al. 2001) and to assess the effects of marsh management on wildlife (Barnby et al. 1985, Collins and Resh 1985, Foin et al. 1997).

The goal of this project was to reconstruct the historical landscape of the Petaluma River watershed prior to major Euro-American modification, and to demonstrate the efficacy of historical hydrology/ecology in identifying and prioritizing multi-benefit restoration opportunities. A large amount of raw data exists pertaining to historical ecological and hydrological conditions within the Petaluma River watershed, some of which has been compiled in previous research (e.g., Heig 1982, Butterworth 1997, Collins et al. 2000), but until now these data have not been synthesized to develop a holistic understanding of the watershed's early landscape. A thorough understanding of historical conditions can be used to guide restoration efforts aimed at restoring fish and wildlife habitat, increasing flood storage and stormwater retention capacity, improving sediment management, enhancing groundwater recharge, and other benefits. The findings from this study can also be integrated with previous research on the historical ecology of other North Bay watersheds—including the Napa River watershed (Grossinger 2012), Novato Creek Baylands (SFEI-ASC 2015b), Laguna de Santa Rosa (Dawson and Sloop 2010, Baumgarten et al. 2017), Sonoma Creek watershed (Dawson et al. 2008), and Miller Creek watershed (Salomon et al. 2008)—contributing to an understanding of historical ecological function and landscape change at a regional scale.

The specific objectives of this study were to:

- Enhance understanding of ecological and hydrological conditions in the Petaluma River watershed prior to major Euro-American modification.

- Develop a GIS map illustrating wetland distribution and channel configuration within the watershed ca. 1850.

- Enhance understanding of the drivers, magnitude, and impacts of landscape change over time.

- Couple findings from the historical landscape research with information on contemporary physical controls to identify potential restoration opportunity areas in the watershed, which will then be further evaluated for restoration feasibility and priority.

- Facilitate communication, education, and outreach to foster a shared understanding of landscape history and a robust dialog around restoration potential.

This project was conducted in collaboration with the Sonoma Resource Conservation District, with funding from the U.S. Environmental Protection Agency.

REPORT STRUCTURE

Chapter 1 provides an overview of the project goals and objectives, as well as the environmental and land use context within the Petaluma River watershed. Chapter 2 describes the methods used in historical data collection, compilation, and synthesis, and in the analysis of landscape change over time. Chapters 3, 4, and 5 describe the habitat types and hydrology that historically characterized the estuary, stream channel network, and non-tidal wetlands, respectively. Chapter 6 assesses the drivers, magnitude, and impacts of landscape changes over time. Chapter 7 discusses potential restoration opportunities throughout the watershed and next steps for research and planning. Scientific names of species referenced in the report are provided in the Appendix.

WHY HISTORICAL ECOLOGY?

Historical ecology is an interdisciplinary field that uses historical data to study ecosystem characteristics (Swetnam et al. 1999, Rhemtulla and Mladenhoff 2007). Through this approach, the typical function and composition of a past landscape is determined by collecting and interpreting the historical materials that documented it—in written accounts and quantitative measurements, cartographic renderings, and photography—before it was significantly altered. Together, the cartographic, photographic, and textual documents provide the converging lines of evidence used to construct an accurate picture of historical landscape patterns.

While urban development, climate change, and other forces have irrevocably altered many aspects of the environment over the past two centuries, many physical controls—such as topography and geology—have remained relatively stable. Historical research can provide relevant clues about how natural, resilient systems persisted in a particular place in the recent past, and how persistent physical controls continue to influence ecological patterns and processes today. As such, historical ecology is a critical component in identifying both the constraints and opportunities posed by the contemporary landscape, and in identifying locally appropriate restoration and management targets (Jackson and Hobbs 2009, Higgs 2012). The study of historical landscapes can also provide clues about how ecosystems were adapted to variable climate regimes, buffering the effects of environmental extremes, which can help with the design of flexible, resilient future ecosystems (Safford et al. 2012). In addition, historical ecology often helps to foster a shared understanding of local landscape history and serves as a valuable educational and communication tool (Hanley et al. 2009).

ENVIRONMENTAL SETTING

The Petaluma River watershed is situated approximately 30 km (20 mi) north of San Francisco, on the northwestern side of San Pablo Bay (Fig. 1.1). The watershed occupies 378 km^2 (146 mi^2), of which 79% is within Sonoma County and the remainder within Marin County. The City of Petaluma, with a population of approximately 60,000, is the main urban area within the watershed. The community of Penngrove is located in the northern portion of the watershed, near Cotati. Urban areas occupy approximately 13% of the watershed, while cultivated or fallow cropland occupies approximately 8% (USDA et al. 2016); dairies, pastureland, and open spaces occupy much of the remainder. Major vegetation communities in the watershed include grassland (15,524 ha; 38,361 ac), montane hardwood forest (3,383 ha; 8,359 ac), coastal salt and brackish marsh (1,995 ha; 4,930 ac), and coast live oak forest/woodland (1,016 ha; 2,511 ac; BAOSC 2017). Tidal wetlands in the watershed experience a semidiurnal tidal regime, with a diurnal tidal range of approximately 1.8–2 m (6–6.6 ft; NOAA tide gages 9415252, 9415423, 9415584).

The watershed is characterized by a Mediterranean climate, with cool, mild winters and warm, dry summers. Precipitation records have been kept at the Petaluma Fire Station for most years since 1872, where annual precipitation averages 60 cm (23 in) over the period of record and 63 cm (25 in) since 1980. Annual precipitation watershed-wide averages about 84 cm (33 in), and varies from less than 20 inches per year in the southeast part of the watershed to about 50 inches per year in higher elevation areas. Most rainfall occurs between the months of November and April (SRCD 2015, WRCC 2016, BAOSC 2017). Summer daily maximum temperatures average 27.6 °C (81.7 °F), while winter daily minimum temperatures average 4.4 °C (40 °F; BAOSC 2017).

Approximately 56% of the watershed is mountainous (SCWA 1986). The Sonoma Mountains form the eastern boundary of the watershed, rising to a maximum elevation of 700 m (2,295 ft) at Sonoma Mountain. The upper elevations of the Sonoma Mountains are comprised of Pliocene-age

Figure 1.1. The Petaluma River watershed encompasses 378 km² (146 mi²) in Sonoma and Marin counties. This study focused on alluvial (valley floor) areas within the watershed (shown in red), which occupy 153 km² (59 mi²). *(NAIP 2016)*

Sonoma Volcanics, which are interbedded at lower elevations by sedimentary rocks of the Petaluma Formation. Several other Neogene volcanics occur within the watershed, including the Burdell Mountain Volcanics on the southern side of the watershed and the Tolay Volcanics around Meacham Hill. Bedrock in the San Antonio Creek drainage on the southern side of the watershed is comprised largely of Jurassic/Cretaceous-age rocks in the Franciscan Complex. The northwestern portion of the watershed, near Wiggins Hill, is dominated by the marine sedimentary rocks of the Wilson Grove Formation. Quaternary alluvial deposits fills the basins and valley floor areas of the watershed, and landslide deposits occupy large areas around Burdell Mountain and the western slopes of Sonoma Mountain. Portions of several faults run through the watershed, including the Petaluma Valley Fault, Tolay Fault, Rodgers Creek Fault, and Burdell Mountain Fault (Langenheim et al. 2010, Wagner and Gutierrez 2010, Wagner and Gutierrez 2017).

The study area for this project includes the alluvial (valley floor) areas throughout the watershed, encompassing a total of 153 km^2 (59 mi^2; see Fig. 1.1).

LAND USE CONTEXT

Archaeological evidence suggests that humans have lived in the Petaluma River watershed for approximately 9,000 years (SRCD 2015). The native Coast Miwok occupied a number of villages throughout the watershed, broadly organized around the Olompali community (in the San Antonio Creek area) and the Petaluma community (centered around the Petaluma River and extending into the Willow Brook, Lynch, and Adobe creek sub-watersheds; Barrett 1908, Milliken 2009). Milliken (2009) estimates that the pre-contact population of these two communities totaled just over 1,000 people. The Coast Miwok utilized a range of natural resources, including fish, shellfish, game, acorns, grasses, and seeds, and managed the landscape in a number of ways, including transplanting California bay trees (Barrett 1908, Stillinger 1982, Lightfoot and Parrish 2009). It is uncertain whether Coast Miwok made regular use of prescribed burning (Lightfoot and Parrish 2009), although an early observer in Sonoma Valley noted that "the grass had been burnt by the Indians of the neighborhood" (Altimira 1823). Coast Miwok communities were disrupted by Spanish colonization during the late 18th and early 19th centuries, and thousands migrated or were forcibly removed to nearby Franciscan missions and Mexican ranchos (particularly Rancho Petaluma; Silliman 2004, Milliken 2009).

From a Creek to a River

The Petaluma River has been referred to as both Petaluma Slough and Petaluma Creek historically. By the mid-19th century, as the City of Petaluma increasingly depended on the waterway for commercial shipping, routine dredging and other channel modifications became necessary to ensure navigability. In 1959, the Petaluma River was officially declared as such by an Act of Congress, qualifying it to receive federal funding for continued maintenance.

The first documented European exploration of the watershed was in 1776, when Fernando Quiros sailed up Petaluma River in search of a water passage to Bodega Bay (Roop and Flynn 2007). Franciscan missionaries explored portions of the watershed in the early 1800s: Luis Arguello and Mariano Payeras traveled through the San Antonio Valley in 1819, and Father Jose Altimira led an expedition through the Petaluma Valley in search of a site for a new mission in 1823. The southern parts of the watershed within and around San Antonio Valley served as grazing lands for the herds at Mission San Rafael between 1817–34, which peaked in size in 1832 (Engelhardt 1897, Bowman 1947). With the disintegration of the mission system in the 1830s, large portions of the watershed were granted to private citizens who submitted land claims to the Mexican government, thus forming the Cotati, Petaluma, Roblar de la Miseria, Laguna de San Antonio, Olompali, and Novato ranchos. Rancho Petaluma, granted to Mariano Vallejo in 1834, occupied almost 18,000 ha (44,000 ac) encompassing the entire eastern portion of Petaluma Valley, and supported cattle herds totaling an estimated 50,000 in the mid-1840s (Silliman 2004).

Large numbers of American settlers first came to the Petaluma area beginning in the 1850s, as the initial wave of Gold Rush prospectors began to

Figure 1.2. Shipping and rail transport enabled the emergence of the City of Petaluma as the region's commercial hub during the mid-19th century. These turn of the century photographs show a freight train crossing over the Petaluma River (b) and scow schooners navigating the river near the railroad drawbridge (a,c). *(a: Photo number B12. 24780, courtesy of San Francisco Maritime National Historical Park; b: Photo number 25049, courtesy of Sonoma Heritage Collection; c: Photo number 25047, courtesy of Sonoma Heritage Collection)*

return from the placers. The Petaluma River was a vital transportation corridor for the region, and the City of Petaluma, incorporated in 1858, rapidly became its primary shipping hub (Fig. 1.2a,c; Munro-Fraser 1880, Heig 1982). As the commercial shipping industry grew, efforts were made to modify the river channel to make it more conducive to maritime navigation. The first attempts to dredge the river were made in 1860, and a major Army Corps project to dredge and straighten the river was initiated in the 1880s (Schulz 1927, Roop and Flynn 2007). Railroad lines were also constructed through the watershed in the late 19th century, including the Petaluma and Haystack Railroad line and portions of the San Francisco & North Pacific Railway Company and Marin and Napa Rail Road Company lines (Fig. 1.2b; Stindt and Dunscomb 1964, Heig 1982, Roop and Flynn 2007).

Agricultural land uses in the watershed were initially dominated by grain cultivation, along with dairying and to a less degree fruit, potato, asparagus, and sugar beet cultivation (*Daily Alta California* 1872, Heig 1982). An 1891 county history, for instance, states that the Petaluma Valley was "until a comparatively recent date devoted to the growing of grain" (Lewis Publishing Company 1891). Thousands of acres of hay and other crops were grown on reclaimed marsh land adjacent to the Petaluma River (see page 73). During the first decades of the 20th century, poultry farming displaced these other land uses to become the dominant agricultural industry in the watershed, and by 1918 Petaluma was known as the "Egg Basket of the World" (Fig. 1.3). By the late 1940s, however, the poultry industry was in decline (Heig 1982).

Highway 101 was constructed through Petaluma in 1956, and by the 1960s urban development had begun to accelerate, with the population increasing from approximately 14,000 to 34,000 between 1960 and 1980 (Fig. 1.4; CSDC 2012). Much of the urban development during the later portion of the 20th century occurred in East Petaluma. The increased population resulted in increased demand for natural resources such as groundwater, and by the 1950s groundwater levels had fallen in many areas, resulting in salt water intrusion into aquifers in the lower part of the valley (Cardwell 1958, SCWA 1986).

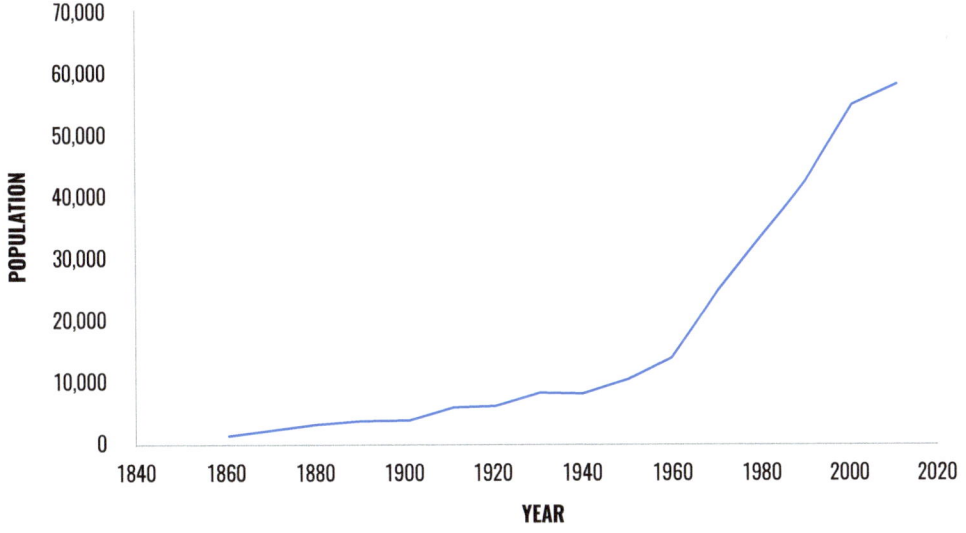

Figure 1.4. Petaluma's population grew rapidly starting in the 1960s, particularly in the eastern portion of the City. *(data from CSDC 2012)*

Figure 1.3. Poultry farming became the dominant industry in Petaluma during the early 20th century. *(Top: Photo number 2013-627-179A, courtesy of Petaluma Historical Library & Museum; Bottom: Photo number 2005-107-89, courtesy of Petaluma Historical Library & Museum)*

2. METHODOLOGY

This chapter details the process used to collect and compile archival data, synthesize the diverse source materials into a historical map, analyze landscape change over time, and identify potential restoration opportunity areas. For additional details on the process used to reconstruct historical landscape characteristics, please consult Grossinger (2005) and Grossinger et al. (2007).

DATA COLLECTION AND COMPILATION

Archival data provided the foundation for the reconstruction of the historical landscape of the Petaluma River watershed. Source materials were collected from 21 local, regional, county, and state archives (Table 2.1), as well as over 20 online databases such as the Online Archive of California, the David Rumsey Map Collection, the California Digital Newspaper Collection, and Jepson Online Journals. The dataset also incorporated a substantial amount of information collected during previous historical ecology research in the Novato Creek and Laguna de Santa Rosa watersheds (Salomon et al. 2015, Baumgarten et al. 2017).

The assembled dataset consists of a variety of historical and contemporary data (Fig. 2.1). Historical materials include maps (e.g., General Land Office [GLO] survey plats, Mexican diseños, U.S. Coast Survey T-sheets and H-sheets, USGS Quads, USDA soil maps, county maps and atlases), photographs (landscape and aerial), and textual documents (e.g., personal diaries, General Land Office field notes, oral history transcripts, technical reports, specimen records). In addition to historical sources, contemporary data in the form of GIS layers (e.g., North Bay LiDAR, SSURGO soil data, modern aerial photos) helped to contextualize and interpret the early landscape. In total, the dataset includes approximately 950 maps, 900 photographs, and 100 pages of transcribed text from roughly 90 different sources.

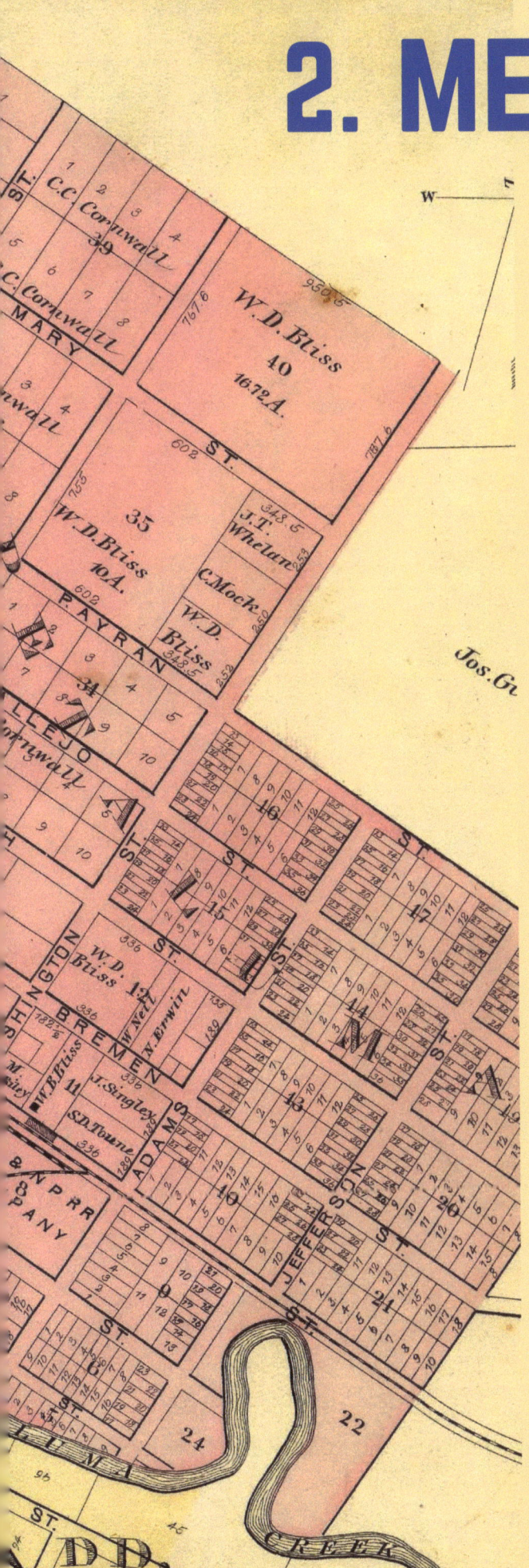

Map of Petaluma and river, 1877. (*Thompson 1877b, courtesy of David Rumsey Map Collection*)

Table 2.1. Source institutions visited or contacted.

Institution	Location
Local and County Archives	
Curtis & Associates, Inc.	Healdsburg
Sonoma State University	Rohnert Park
Petaluma Regional Library Petaluma History Room	Petaluma
Sonoma County History and Genealogy Library	Santa Rosa
Petaluma Historical Library and Museum	Petaluma
Sonoma County Surveyor	Santa Rosa
Sonoma County Recorder/Assessor	Santa Rosa
Marin County Recorder/Assessor	San Rafael
Marin County Free Library Anne T. Kent California Room	San Rafael
Bay Area Regional Archives	
The Bancroft Library, UC Berkeley	Berkeley
UC Berkeley Earth Science and Map Library	Berkeley
UC Berkeley Bioscience and Natural Resources Library	Berkeley
California Historical Society	San Francisco
Society of California Pioneers	San Francisco
San Francisco Maritime National Historical Park Research Center	San Francisco
State Archives	
California State Archives	Sacramento
California State Library	Sacramento
State Lands Commission	Sacramento
Other	
Bureau of Land Management	Sacramento
University of Alabama Map Library	Tuscaloosa, AL
Water Resources Collection & Archives, UC Riverside	Riverside

Historical sources that were spatial in nature, captured a high degree of landscape detail, or depicted unique features (maps and spatially explicit narrative data) were selected from the dataset, compiled, and georeferenced in a Geographic Information System (GIS) database. Approximately 50 historical maps, as well as coordinate-based point data such as GLO field notes and species observations from Consortium of California Herbaria (CCH) and VertNet databases, were georeferenced.

Historical aerial photographs of the Petaluma River watershed (Fig. 2.2), composed of over 120 individual frames from a 1942 survey (USDA 1942), were orthorectified and mosaicked by SFEI to create a continuous image of the study area.

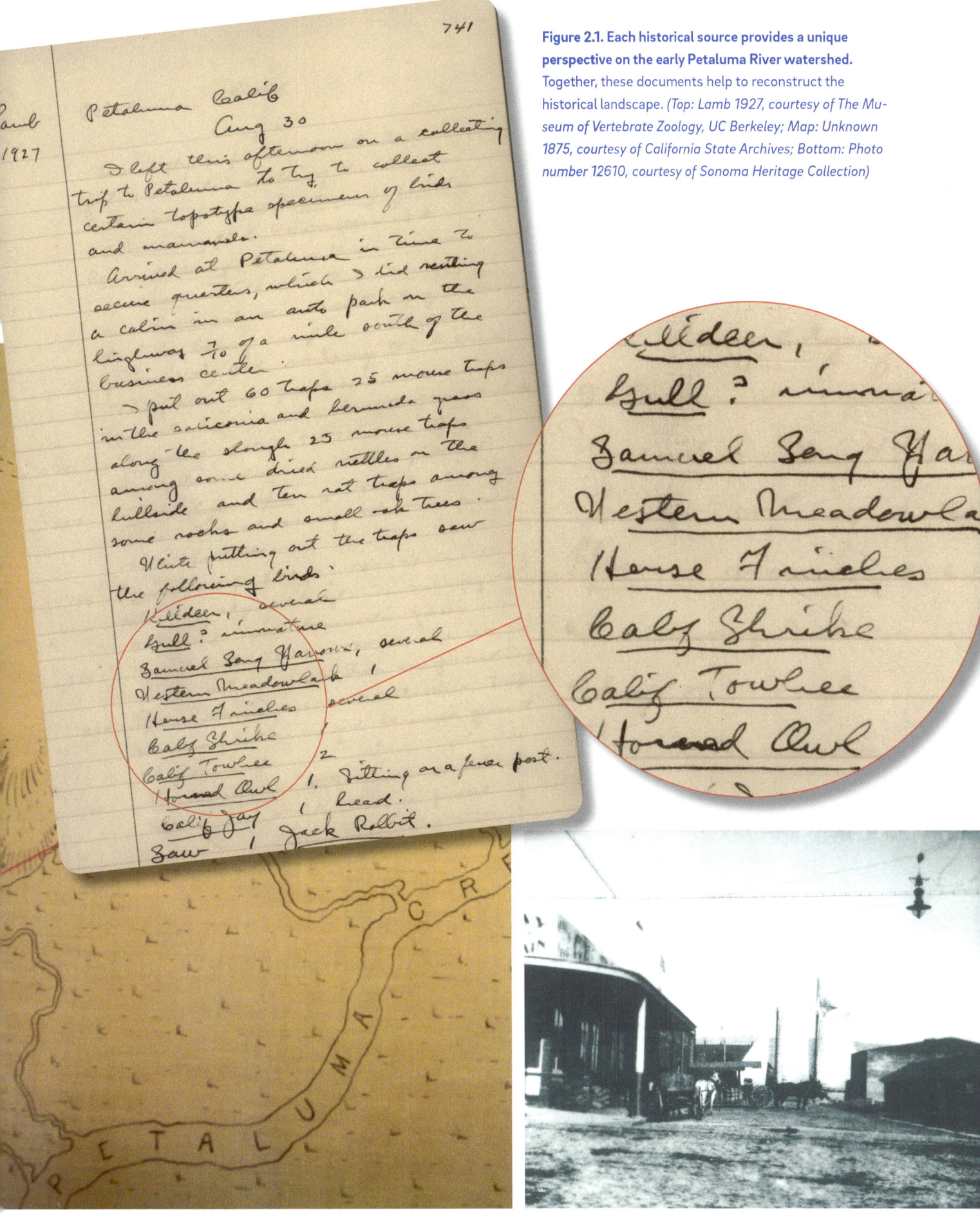

Figure 2.1. Each historical source provides a unique perspective on the early Petaluma River watershed. Together, these documents help to reconstruct the historical landscape. *(Top: Lamb 1927, courtesy of The Museum of Vertebrate Zoology, UC Berkeley; Map: Unknown 1875, courtesy of California State Archives; Bottom: Photo number 12610, courtesy of Sonoma Heritage Collection)*

PETALUMA VALLEY HISTORICAL HYDROLOGY AND ECOLOGY STUDY

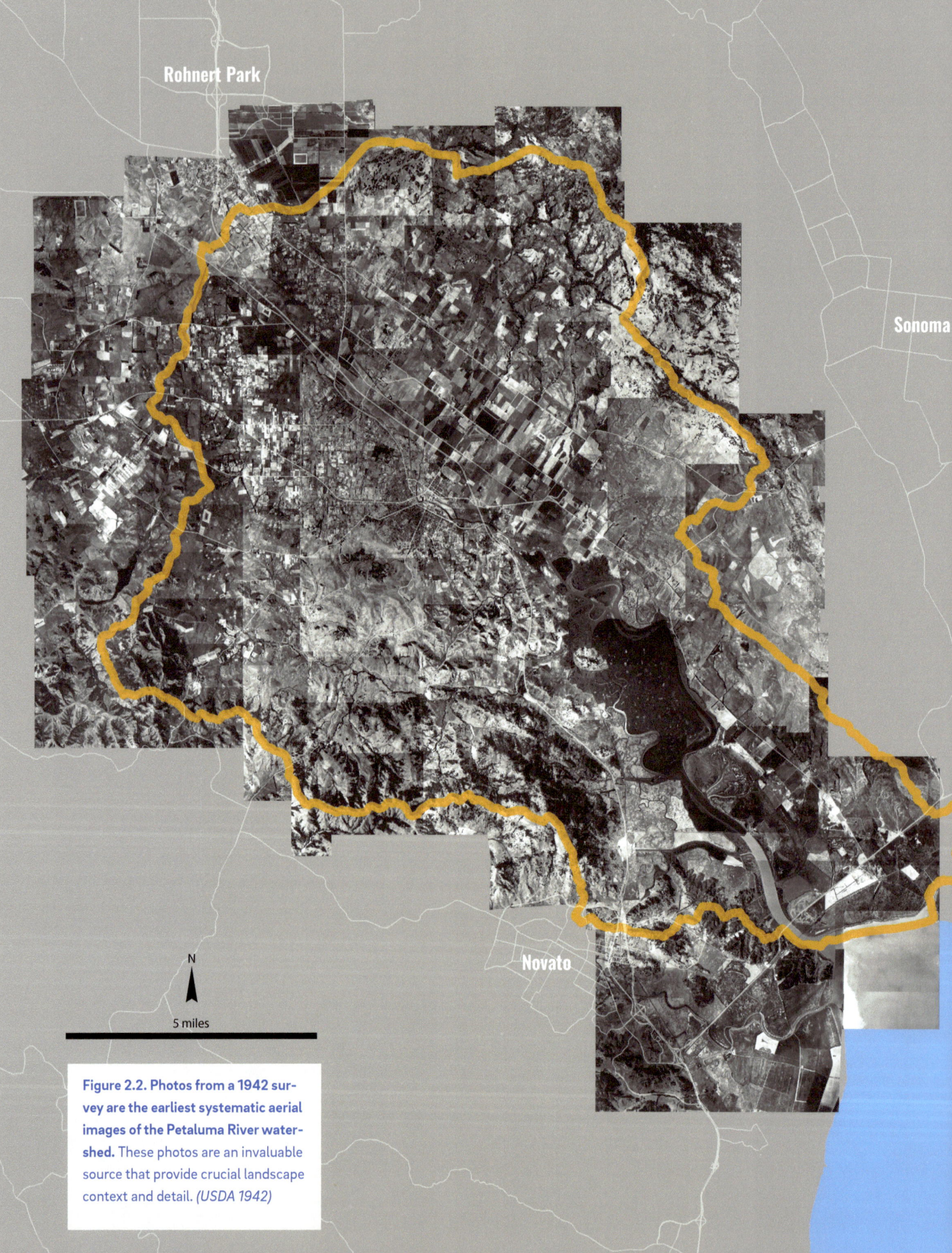

Figure 2.2. Photos from a 1942 survey are the earliest systematic aerial images of the Petaluma River watershed. These photos are an invaluable source that provide crucial landscape context and detail. *(USDA 1942)*

GIS SYNTHESIS AND MAPPING

Two GIS layers, representing the historical distribution of mid-19th century wetland habitat types and the configuration of stream networks, were developed through the synthesis and interpretation of compiled data (Table 2.2). Eight historical wetland habitat types (four tidal and four non-tidal) were defined and mapped as polygon features, and four types of historical channels were mapped as line features (one channel type was mapped as a polygon feature). A forked distributary symbol was used to indicate the locations where channels terminated historically. The habitat and channel features were digitized according to the earliest and most reliable sources (often the 1942 aerial photos, T-sheets, and USGS Quads). The boundaries of certain features were digitized from multiple sources in order to capture the most representative shape and location. The digitizing sources for each feature, or segment of a feature, were recorded in GIS attribute tables. A number of supporting materials were also referenced to develop an understanding of, and context for, the historical features. These interpretation sources were also documented in the attribute tables. Finally, each feature was assigned a high, medium, or low certainty level for three metrics of accuracy (interpretation/classification, size/shape, and location).

Habitat polygons and channel features were mapped from the most spatially accurate sources believed to be representative of historical landscape condition and configuration. Tidal wetlands were generally digitized from early United States Coast Survey topographic sheets (T-sheets; see pages 24-25). Where T-sheet coverage was not available (around Rush Creek), tidal wetlands were digitized from other early sources such as General Land Office survey plats or 1942 aerial photographs; mapping detail is lower in areas where T-sheet coverage was not available. In many other cases, wetland and channel features were digitized from either the 1942 aerial photographs (USDA 1942), the modern LiDAR-derived DEM (County of Marin 2015), or modern soils data. Wherever possible,

Table 2.2. Historical wetland and channel types included in GIS synthesis mapping.

	Wetland/Channel Type	Description
Tidal	Tidal Marsh	Wetland plain with a salinity gradient that decreases with distance from San Pablo Bay
	Tidal Mudflat	The zone of the channel in between tide marks; above water during low tide and submerged at high tide
	Subtidal Channel	The zone of the channel submerged except during extremely low tides. (Shown together with Bay in the historical and modern mapping.)
	Marsh Panne/Pond	Shallow depressions that form within the marsh plain and often develop higher salinities than the surrounding marsh
Non-tidal	Wet Meadow	Seasonal wetland type often associated with clay dominated soils; dominated by herbaceous plants such as grasses, rushes, and forbs
	Valley Freshwater Marsh	Perennial wetland type dominated by tules and cattails
	Vernal Pool Complex	Seasonal depressional wetland type covered by shallow water for a period between winter and spring, and dry for the rest of the year. Features may be connected to each other by swales, forming complexes
	Willow Grove	Often occur at the ends of channels or on alluvial fans where there is emergent groundwater; typically dominated by arroyo willow
Channels	Perennial Channel	Stream that supports year-round flow
	Intermittent Channel	Stream that flows seasonally
	Slough	Emergent on the valley floors; conveyed flows through seasonal wetland complexes
	Small Tidal Flat	Small channels with the marsh plain that are inundated at high tide and return to mudflat at ebb tide. (Shown together with Tidal Mudflat in the historical mapping.)
	Large Pool (in-channel)	Deep pool within the stream channel

early sources (i.e., mid-late 19th century maps and textual data) were used to confirm the historical presence of a particular feature and establish its approximate shape, size, location, and classification.

ANALYSIS OF CHANGE OVER TIME

A map of contemporary wetland habitat types was developed in order to compare and analyze changes in habitat distribution and channel configuration over time. The contemporary map was compiled primarily from the Bay Area Aquatic Resources Inventory (BAARI; SFEI-ASC 2015a) as well as additional features from the Sonoma County Fine Scale Vegetation and Habitat Map data layers (Sonoma Veg Map 2017). Modern classifications were then crosswalked to historical habitat types (Table 2.3). Three novel modern habitat classifications were included in the change analysis to represent habitat types that are not comparable to any historical habitat types, but have developed on the modern landscape: 1) Unnatural Lagoons (artificial impoundments of water; may be tidal or non-tidal, vegetated or unvegetated), 2) Depressional Vegetated (vegetated topographic depressions that receive water primarily from precipitation and overland flow), and 3) Depressional Open Water (topographic depressions that receive water primarily from precipitation and overland flow).[1]

Landscape change was determined by comparing the relative extent or length of historical features to the modern map. Analysis was limited to habitats within the alluvial boundary.

---

[1] Some of the contemporary Depressional Vegetated wetlands could be Valley Freshwater Marshes comparable to historical wetlands of this type, but we did not have sufficient data to classify them as such.

Table 2.3. Crosswalk between modern and historical habitat types. Modern wetland mapping was compiled primarily from the Bay Area Aquatic Resources Inventory (BAARI; SFEI-ASC 2015a); vernal pool mapping is derived from the Sonoma County Fine Scale Vegetation and Habitat Map data layers (Sonoma Veg Map 2017).

Source	Modern Classification	Crosswalk Classification
BAARI Baylands	Lagoon Perennial Open Water Unnatural	Unnatural Lagoon
BAARI Baylands	Lagoon Perennial Unvegetated Flat Unnatural	Unnatural Lagoon
BAARI Baylands	Lagoon Perennial Vegetated Unnatural	Unnatural Lagoon
BAARI Baylands	Tidal Vegetation	Tidal Marsh
BAARI Baylands	Shallow Bay	Subtidal Channel and Bay
BAARI Baylands	Tidal Bay Flat	Tidal Mudflat
BAARI Baylands	Tidal Ditch	Tidal Mudflat
BAARI Baylands	Tidal Marsh Flat	Tidal Mudflat
BAARI Baylands	Tidal Panne	Marsh Panne/Pond
BAARI Wetlands	Depressional Vegetated Natural	Depressional Vegetated
BAARI Wetlands	Depressional Vegetated Unnatural	Depressional Vegetated
BAARI Wetlands	Playa Unvegetated Flat Unnatural	Marsh Panne/Pond
BAARI Wetlands	Playa Vegetated Unnatural	Unnatural Lagoon
BAARI Wetlands	Seep or Spring Natural	Wet Meadow
BAARI Wetlands	Seeps or Spring Unnatural	Wet Meadow
BAARI Wetlands	Depressional Open Water Natural	Marsh Panne/Pond
BAARI Wetlands	Depressional Open Water Unnatural	Depressional Open Water
Sonoma Veg Map	Western North America Vernal Pool Macrogroup	Vernal Pool Complex

IDENTIFICATION OF POTENTIAL RESTORATION OPPORTUNITY AREAS

Findings from the historical reconstruction and landscape change analyses, combined with a preliminary analysis of contemporary land uses and physical controls, were used to identify potential opportunity areas for restoring a range of wetland and riparian habitat types throughout the watershed, including tidal marshes, tidal-terrestrial transition zones, freshwater wetlands, riparian forests, and riparian wetlands. The analysis was limited to open spaces where wetlands currently do not exist.[2]

Open spaces (i.e., undeveloped areas, protected areas, and agricultural lands) were identified using Sonoma County Fine Scale Vegetation and Habitat Map data layers (for Sonoma County; Sonoma Veg Map 2017) and the 2016 USDA Cropland Data Layer (for Marin County; USDA 2016). NRCS SSURGO soil data (NRCS 2011, 2013) was used to identify areas with "somewhat poorly drained," "poorly drained," or "very poorly drained" soils within the open spaces, which were classified as potential freshwater wetland restoration opportunity areas. Special Flood Hazard Area layers from the FEMA NFHL dataset (FEMA 2016, 2017) were used to identify areas within the 100-year floodplain within the open space areas, which were classified as potential riparian forest restoration opportunity areas. Open space areas that are both within the FEMA 100-year floodplain and in areas with poorly drained soils were classified as potential riparian wetland restoration opportunity areas.

NOAA Coastal Services Center Sea Level Rise Inundation Data (NOAA 2012a,b) were used to identify potential tidal marsh and transition zone restoration opportunity areas. Areas at or below current Mean Higher High Water (MHHW) were classified as potential tidal marsh restoration opportunity areas. Areas between current MHHW and MHHW + 2 m (6 ft) were classified as potential transition zone restoration opportunity areas; these could also include areas suitable for marsh migration under future sea level rise scenarios. Potential tidal marsh and transition zone restoration opportunity areas also include some areas within the FEMA 100-year floodplain and/or areas with poorly drained soils.

Further feasibility analysis, taking into account factors such as land ownership, landowner interest, soil quality, groundwater elevations, and historical land cover, will be needed to identify and prioritize sites appropriate for restoration activities (see Chapter 7).

TRANSFERABILITY OF METHODS

Over the past 25 years, SFEI's historical ecology research has informed the development of ecosystem management and restoration strategies throughout the Bay Area, Central Valley, and coastal California. Drawing on a largely untapped dataset, historical ecology offers a unique perspective that can alter our understanding of how landscapes function and reveal previously unrecognized restoration opportunities. Insights from historical ecology research can help address a wide range community concerns around environmental stewardship, flood control and sediment management, climate change adaptation, and other issues.

The methodology used in this study is broadly transferable to many other settings statewide, particularly the valley floor portions of watersheds in agricultural settings. Key sources used to examine the historical hydrology and ecology of the Petaluma River watershed, such as General Land Office survey notes and plats, soil maps, land grant case maps, and early aerial photographs, are

---

[2] Because existing non-tidal wetlands may become suitable sites for transition zone restoration with future sea level rise, existing wetlands were included in the potential transition zone restoration opportunity areas.

available in many areas of the state. While these key sources alone did not provide a complete picture of the historical landscape within the study area, they are spatially explicit and relatively comprehensive in their coverage, and for most habitat types they provided sufficient information to enable a coarse reconstruction of historical ecological patterns.

The historical aerial photographs were particularly useful for mapping channels and seasonal wetlands, though they were produced relatively late and required a substantial effort to orthorectify and mosaic. Historical and contemporary soil maps, used in combination, were essential in reconstructing the distribution of freshwater wetlands. GLO survey and land grant maps provided valuable information about riparian forest composition, freshwater wetland features, degree of tidal influence, channel characteristics, and other topics. Coast Survey maps, the primary source used to reconstruct estuarine habitat types, are only available for coastal locations.

A wide variety of supplemental sources, such as parcel and subdivision maps, landscape photographs, botanical records, oral histories, explorer diaries, county histories, and travelogues were used to refine and verify the basic ecological patterns documented in key sources. These sources provided information about vegetation composition, wildlife support, channel configuration, flood extent and frequency, and land use history, and greatly increased the specificity and complexity associated with the historical reconstruction. While most of these sources are likely to be available in many other regions of the state, their depictions or descriptions of the historical landscape tend to be incomplete and somewhat idiosyncratic, and thus they may not provide useful ecological information for a particular site.

TECHNICAL ADVISORY COMMITTEE

Research for this study was conducted with the support and guidance of a technical advisory committee (TAC) composed of 18 local experts (Table 2.4). TAC members contributed to the development of project objectives, methodology, analysis, interpretation, and reporting. Two meetings (January 2016 and June 2017) were held with the TAC to present preliminary findings and solicit advisors' comments and feedback; advisors also provided comments on a draft version of this report.

Table 2.4. Technical Advisory Committee members. Acronyms: Resource Conservation District (RCD), Friends of the Petaluma River (FOPR), Sonoma County Water Agency (SCWA), National Oceanic and Atmospheric Administration (NOAA), California Department of Fish and Wildlife (CDFW), Sonoma Land Trust (SLT), Natural Resources Conservation Service (NRCS), North Bay Watershed Association (NBWA), Environmental Protection Agency (EPA), Environmental Science Associates (ESA).

Advisor	Affiliation
Betty Andrews	ESA
Stephanie Bastianon	FOPR
Jason Beatty	City of Petaluma
Laurel Collins	Watershed Sciences
John Fitzgerald	Retired engineer
Susan Haydon	SCWA
Judy Kelly	NBWA
John McKeon	NOAA
Julian Meisler	SLT
Andy Rodgers	FOPR
Solange Russek	Petaluma Historical Library & Museum
Nancy Scolari	Marin RCD
Gail Seymour	CDFW
William Stockard	Sonoma County
Chase Takajo	SCWA
Jared Vollmer	EPA
Jennifer Walser	NRCS
Jason White	ESA

3. ESTUARY

Drawing of Petaluma waterfront (undated). *(Photo number 2013-627-95A, courtesy of Petaluma Historical Library & Museum)*

OVERVIEW

Historically (ca. 1850), the lower Petaluma River meandered through a vast area of tidal wetlands on the northwest side of San Pablo Bay. Like other tidal marshes in the San Francisco Estuary, the Petaluma Marsh formed over thousands of years as sea levels began to rise at the end of the last glacial period 10,000 to 12,000 years ago (Malamud-Roam and Goman 2012). Initially, the rising tides rapidly filled valleys and coastal wetlands (at a rate of ~6–8 mm/year; 0.24–0.3 in/year), before slowing around 6,000 years ago (to ~1–2 mm/year; 0.04–0.08 in/year). The oldest tidal marshes in the San Francisco Estuary, including the Petaluma Marsh, began to form during this period of decelerating sea level rise (Malamud-Roam and Goman 2012). Colonizing plant species that had migrated upstream with the rising seas facilitated the upward growth of the marsh plain, in part by retaining sediment and accumulating organic matter. The estuary's mixed semidiurnal tides and its seasonal freshwater inputs from the Petaluma River produced a pattern of salinity variation that was reflected in the eventual distribution of plant communities. Additionally, the dense, complex channel network acted as an important control on salinity and inundation locally, which strongly structured plant composition according to the size and proximity of tidal channels (Byrne et al. 1998, Sanderson et al. 2001, Sanderson et al. 2000, Watson and Byrne 2009, Watson 2012).

Tidal wetlands occupied 6,540 ha (16,150 ac) in the Petaluma River watershed during the mid-19th century, and included 4,570 ha (11,290 ac) of tidal marsh, 590 ha (1,460 ac) of subtidal channels, 1,250 ha (3,090 ac) of tidal mudflats and sloughs, and 120 ha (310 ac) of marsh ponds and pannes. The marsh extended 20 km (12 mi) to the northwest of San Pablo Bay, and, at its widest point, spanned roughly 11 km (7 mi). The Petaluma River was the central feature of the historical estuary. The river's course was narrow and highly sinuous in the upper estuary and widened as it approached its mouth at the bay. The mainstem was surrounded by a broad, fan-shaped marsh plain that tapered to a point in the upper estuary. Several large sloughs—Black John, San Antonio, Schultz and Tule—branched off of the river's mainstem, and in the lower estuary a large intertidal feature known as False Bay adjoined the main channel. The estuary received both freshwater and sediment from the fluvial reach upstream and from its largest tributary, San Antonio Creek, which connected to the tidal portion of the river via San Antonio Slough east of Burdell Island. Tidal influence extended from San Pablo Bay more than 20 km (12 mi) inland to the head of tide near the confluence with Lynch Creek (Fisher 1852).

United States Coast Survey topographic sheets (T-sheets), hydrographic sheets (H-sheets), and composite charts provide some of the earliest and most reliable depictions of the morphology of the historical estuary (Fig. 3.1). Two T-sheets (T-817 and T-818) that cover the majority of the estuary were surveyed by Augustus Rodgers and David Kerr in 1860. Kerr, in particular, is recognized for his meticulous technique and extraordinarily detailed drawings that consistently captured even minor sloughs and pannes (Grossinger et al. 2005). Two other T-sheets that cover the mouth of the Petaluma River at San Pablo Bay and portions of the surrounding marsh were surveyed by Rodgers in 1854 (T-472) and 1856 (T-564). Two H-sheets (H-724 and H-725), surveyed by James Alden in 1860, show channel depths within the tidal portion of the Petaluma River. The precision of these early documents make them an invaluable source for representing and quantifying metrics of the early landscape, as well as contextualizing other historical documents that depict the estuary.

This chapter summarizes the historical evidence for the hydrological and geomorphic characteristics of the river and tidal channel network, the composition and distribution of marsh habitats and vegetation, and the species support functions provided by the historical estuarine landscape.

Figure 3.1. U.S. Coast Survey charts depict the mid-19th century Petaluma River estuary in rich detail. Surveyors Augustus Rodgers, David Kerr, and James Alden charted tidal habitats, channel networks, and channel depths directly from the field to render highly spatially accurate maps. Early topographic sheets (T-sheets) and hydrographic sheets (H-sheets) created from these surveys provided a crucial foundational picture of the historical Petaluma River estuary; the chart shown here is a simplified composite map derived from multiple surveys. Note that several portions of the estuary, particularly around Rush Creek, were not included in the surveys. *(USCS 1861, courtesy of NOAA)*

PETALUMA RIVER AND TIDAL CHANNELS

From its interface with the fluvial reach upstream, the Petaluma River meandered along its tidal course towards San Pablo Bay. The tidal reach measured approximately 28 km (17 mi) along the mainstem, and was characterized by a high degree of sinuosity, especially in the upper third of the estuary. In the lower estuary, several large sloughs branched off of the mainstem, including Black John Slough, which joined Rush Creek in a southwestern lobe of the marsh, San Antonio Slough, which connected to the creek that flowed through Chileno Valley in the west, Schultz Slough, which branched off the mainstem north of Neil's Island, and Tule Slough, which branched off the mainstem east of Neil's Island.

The Petaluma River was subtidal for most of its length through the estuary, with large margins of intertidal flats near its mouth and at False Bay. The 1860 USCS H-sheets recorded a depth of 4 m (12 ft) at the San Pablo Bay inlet, which measured approximately 1,080 m (3,540 ft) wide (Fig. 3.2). A maximum depth of 13 m (44 ft) was recorded in the channel on the north side of Hog Island (Alden 1860a). Depths navigable for shallow-draught ships continued up to Haystack, with pockets of deeper water in the channel bends (*Daily Alta California* 1872; Fig. 3.3). The width of the subtidal channel narrowed from roughly 600 m (1969 ft) above the inlet to about 135 m (443 ft) at Lakeville, 34 m (112 ft) near Haystack, and up to about 20 m (66 ft) within the Petaluma city limits (Alden 1860b).

One of the most distinctive features of the tidal portion of the Petaluma River was its sinuosity in the upper estuary. The river's meanders were likened to an "everlasting corkscrew—running two or three miles to gain one" and presenting early mariners with a navigational hazard (*Daily Alta California* 1872, Panoramics 1860). A ca. 1880 landscape photo looking north into the city of Petaluma captured these meanders, so protracted and circuitous that they drift in and out of the frame (Fig. 3.4). As a mid-19th century observer remarked:

> *As we approached Black Point, at the mouth of Petaluma Creek, the water of the bay became very shallow and muddy, and our course changed from a right line into a tortuous following of the narrow channel...The creek, which is a mere tide-water slough, winds its labyrinthine way through an expanse of reedy marsh. (Taylor 1862)*

The river's sinuous morphology resulted from a combination of fluvial and tidal processes that included large freshwater and sediment inputs during flooding events, as well as tidal action that circulated suspended sediment and scouring currents throughout the channel network. The spatial and temporal variability of these inputs contributed to the distinct morphology of the tidal reach. Major flood events during the wet season (see pages 39–42) conveyed large volumes of freshwater from the fluvial reach towards the marsh. During an 1899 flood, for instance, the *Press Democrat* described the marsh south of the city as "one vast sheet of water" (*Press Democrat* 1899). Much of the sediment transported during these flood events was eventually deposited in the marsh below. After a period of heavy rainfall in February 1869, for instance, sediment that had washed down from

---

Its course very much resembles the track of a man who has spent half an hour hunting for a lost pocketbook in a field. If, after gazing awhile at the creek, the eye should suddenly be turned to a ram's horn or a manzanita stick, the latter would appear perfectly straight, by comparison.

–Hutching's California Magazine 1859 in Roop and Flynn 2007, describing the Petaluma River

Figure 3.2. The channel geometry of the Petaluma River, recorded in feet and fathoms, was surveyed by the United States Coast Survey in 1860. This map shows a depth of 4 m (12 ft) at the San Pablo Bay inlet. Hydrographic sheets like this one were crucial navigational tools for mariners sailing up the meandering tidal channel. *(Alden 1860a, courtesy of NOAA)*

Figure 3.3. A schooner carrying a load of hay embarks on the Petaluma River (no date). The Petaluma River was navigable at high tide to Haystack, about 3 km (2 mi) south of the city of Petaluma. *(Photo number 2005-273-84, courtesy of Petaluma Historical Library & Museum)*

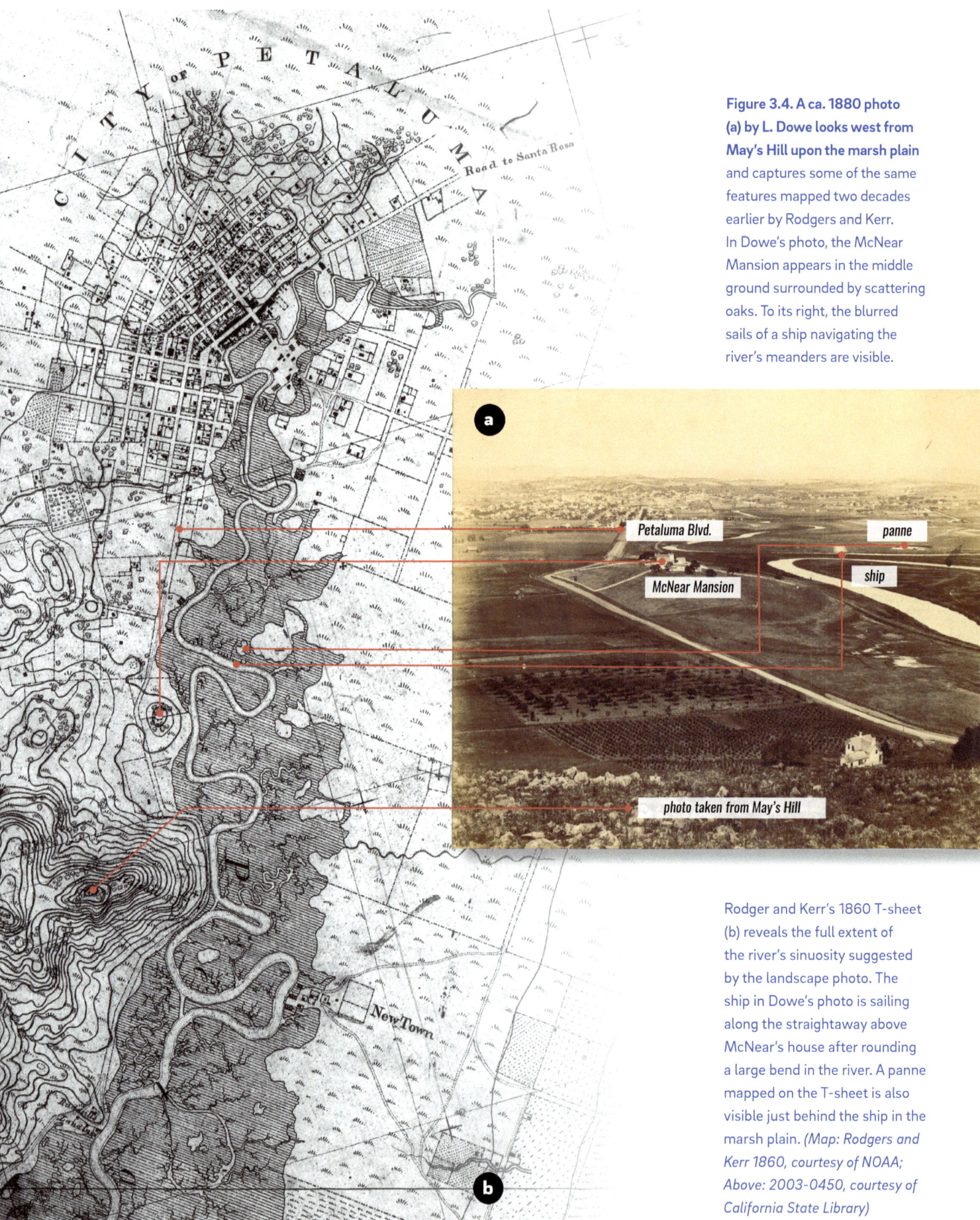

Figure 3.4. A ca. 1880 photo (a) by L. Dowe looks west from May's Hill upon the marsh plain and captures some of the same features mapped two decades earlier by Rodgers and Kerr. In Dowe's photo, the McNear Mansion appears in the middle ground surrounded by scattering oaks. To its right, the blurred sails of a ship navigating the river's meanders are visible.

Rodger and Kerr's 1860 T-sheet (b) reveals the full extent of the river's sinuosity suggested by the landscape photo. The ship in Dowe's photo is sailing along the straightaway above McNear's house after rounding a large bend in the river. A panne mapped on the T-sheet is also visible just behind the ship in the marsh plain. *(Map: Rodgers and Kerr 1860, courtesy of NOAA; Above: 2003-0450, courtesy of California State Library)*

the hills was reported to have been deposited in the marsh ankle-deep in places and over knee-deep in others (Parker 1869).

In addition to fluvial inputs, tidal flux also transported water and suspended sediment throughout the marsh. Tidal influence varied spatially within the estuary, and decreased with distance from the tidal source. Near the mouth of the Petaluma River, contemporary research has documented a repeating process of sediment erosion and deposition, resulting in an oscillating plume of sediment that moves back and forth between the Petaluma River and San Pablo Bay, depositing sediment at slack tides and re-suspending it during ebb tides (Schoellhamer 2003). Within the marsh, most sediment is deposited near channels. In addition, the sediment concentration of water reaching first-order channels is much lower than the concentration along higher order channels closer to the mouth of the estuary (Collins et al. 1987). In general, the Petaluma Marsh's elongated configuration along the valley shelters it from wave energy and limits the influence of wave-driven erosion or sediment delivery (Goals Project 2015).

Aside from the mainstem Petaluma River and several large sloughs, the majority of channels in the estuary were intertidal and formed a dense dendritic network throughout the marsh (Fig. 3.5). A few of the large sloughs that branched off the mainstem—Black John, San Antonio, and Schultz—were deep enough to have some subtidal water, but for the most part, the tidal channels throughout the marsh became mud flats at low tide. These channels were highly sinuous and branched off into numerous smaller channels that transported sediment and nutrients throughout the marsh. In total, there were approximately 440 km (270 mi) of small, intertidal channels in the historical marsh.

Figure 3.5. **Highly detailed mapping by USCS surveyors captures the dense tidal channel network** throughout the marsh plain. Tidal sloughs branched off into small channels that circulated the tides throughout the estuary. *(Kerr 1860, courtesy of NOAA)*

MARSH PLAIN

The Petaluma Marsh historically measured approximately 11 km (4 mi) across from east-to-west at its widest point, and tapered to around 74 m (243 ft) wide in the upper estuary near present-day Petaluma. The marsh included a large lobe on its southwestern side where Rush Creek flowed from the northern boundary of Novato into the Petaluma River. A large shallow-water feature known as False Bay adjoined the marsh on the east. A short subtidal channel reached into False Bay south of Hog Island, but the majority of the bay (2 km [1 mi] long, 1 km [0.5 mi] wide) was intertidal. The marsh plain that surrounded the channel network contained a mosaic of habitats that included numerous pannes and heterogeneous vegetation communities. The marsh also contained a dense, branching tidal channel network that circulated the tides throughout the marsh system via thousands of small intertidal channels (see Fig. 3.5).

Historically, the Petaluma Marsh exhibited a high degree of vegetational heterogeneity. Early descriptions of the marsh often remark on the vast tule plains that dominated the landscape (Loring 1852, Menefee 1873, Smith 1906, Cardwell 1958, Panoramics 1860). However, these early references to tules are likely somewhat of a misrepresentation, as "tules" was often used as a generic description for marsh vegetation. The surveyors who traversed the marsh may have also over-represented the species, since their routes frequently corresponded to tule habitat along shorelines and tidal sloughs. In addition to tules, numerous records of pickleweed were documented, particularly along the edges of sloughs (Dixon 1908 and 1909, Lamb 1927). Early specimen records also list California cordgrass, fleshy jaumea, alkali heath, hairy gumweed, salt grass, sea beet, alkali bulrush, sea milkwort, and soft bird's beak, among other marsh plant species (Table 3.1).

Across the marsh plain, the spatial distribution of vegetation was influenced by multiple factors that include proximity and size of tidal channels, climate, salinity, and elevation. Sanderson et al. (2000) showed that the distribution and richness of vegetation in Petaluma Marsh is strongly associated with distance to tidal channels of different sizes, with an average of 1.6 additional species found within 10 m (33 ft) of channels. Their results suggest that even the smallest channels (50 cm [20 in] wide and 75 cm [30 in] deep) can influence the distribution of both major and minor species. Sanderson et al. (2001) found that, in addition to size, channel geometry was an important control, with greater influence on distribution inside channel bends than outside them. The richness of marsh species has also been shown to be negatively correlated with soil salinity (Vasey et al. 2012).

An abundance of ponds and pannes, cumulatively occupying 120 ha (310 ac), were found in the Petaluma Marsh plain (Fig. 3.6). These pannes are characteristic of the San Francisco Estuary, but are "relatively infrequent in other central coast tidal marshes" (Baye et al. 2000). Panne formation can result from a number of mechanisms, including sparse vegetation cover, channel retrogression, low peat production, isolation from fluvial influences, differential rates of primary production, and low effective precipitation. When an initial depression forms within the marsh vegetation, the shallow panne retains water that eventually develops a salinity greater than the surrounding environment, further inhibiting growth of vegetation and contributing to panne formation (Collins et al. 1987, Goudie ed. 2004, Malamud-Roam and Goman 2012). These pannes provided microhabitats within the Petaluma Marsh plain that supported specialized communities of plants and animals; characteristic plant species found in marsh ponds/pannes may have included widgeongrass, salt marsh sand spurry, and alkali heath (Koch 1940, Baye et al. 2000, CNPS 2017; see Table 3.1).

Table 3.1. **Partial list of native plants associated with the estuarine landscape that were historically present in the Petaluma River watershed,** drawn from herbarium records. Where available, relevant information about the locality where the specimens were collected is included. Species were selected for inclusion in the table based on a combination of the locality information provided in the herbarium records and known associations with estuarine types. All data were provided by the participants of the Consortium of California Herbaria.

Common Name	Scientific Name	Relevant Excerpt(s)	Year(s)	Notes
Alkali bulrush	Bolboschoenus maritimus	Burdell Sta., "Petaluma," "Salt-marsh"	1897, 1945	
Alkali heath	Frankenia salina	Petaluma marshes	1930	
Alkali plantain	Plantago maritima var. juncoides	San Antonio, "Salt marshes"	1945	
Arrowgrass	Triglochin maritima	Petaluma, "Salt marshes"	1880	
Baltic rush	Juncus balticus	Petaluma marshes	1897	
California aster	Symphyotrichum chilense	Petaluma	1880, 1882	
California cordgrass	Spartina foliosa	Along Petaluma Creek, below drawbridge, 1 mile northeast of Black Point, "Burdell marshes," "Tidal flats"	1880, 1927, 1945	
Canyon dodder	Cuscuta subinclusa	Burdell Station, "Mouth of San Antonio Creek," "Salt marsh"	1939, 1945	Listed in records as Cuscuta ceanothi.
Dwarf peppergrass	Lepidium latipes	Petaluma	1921	
Fleshy jaumea	Jaumea carnosa	Petaluma	1880	
Fowler's knotweed	Polygonum fowleri	Burdell Station, "Salt marsh"	1945	
Hairy gumweed	Grindelia hirsutula	Petaluma	1880, 1881, 1930	
Marin knotweed	Polygonum marinense	Burdell; salt marshes	1945	
Olney's three-square bulrush	Schoenoplectus americanus	Burdell, "Petaluma (salt-marsh)"	1897, 1945	Listed in some records as Scirpus americanus.
Petaluma popcornflower	Allocarya vestita	Petaluma	1880	CNPS Rare Plant Rank 1A. Collection date appears to be mislabeled as 1850.
Salt grass	Distichlis spicata	In salt marshes	1897	
Salt marsh sand spurry	Spergularia salina	Petaluma	1921	
Saltmarsh dodder	Cuscuta salina var. major	Mouth of San Antonio Creek, "Salt marsh"	1939	
Sand spurry	Spergularia macrotheca var. macrotheca	Petaluma	1921	
Sea beet	Beta vulgaris	Petaluma, "Salt-marsh"	1897	
Sea milkwort	Lysimachia maritima	Petaluma, "Salt marsh"	1897	Listed in record as Glaux maritima.
Smooth tidy tips	Layia chrysanthemoides	Petaluma	1921	
Soft bird's-beak	Chloropyron molle subsp. molle	Burdell Station, "Salt marsh"	1897, 1945	CNPS Rare Plant Rank 1B.2. Listed in some records as Cordylanthus mollis subsp. mollis.
Suisun marsh aster	Symphyotrichum lentum	Petaluma Marshes, near rail road track	1897	CNPS Rare Plant Rank 1B.2. Listed in record as Aster lentus.
Suisun marsh aster	Symphyotrichum lentum	Near Petaluma	1888	CNPS Rare Plant Rank 1B.2.
Widgeongrass	Ruppia maritima	Petaluma, "In water of saline flat," "In shallow tidal ditches"	1897, 1940	

Figure 3.6. The early T-sheet surveys (a) documented 733 marsh pannes, ranging in size from 9–35,420 m² (97–10,068,770 ft²). These pannes contributed to marsh heterogeneity and complexity and provided important habitat for wildlife and vegetation. A 1900 photo (b) looking south on the Petaluma Marsh from Haystack Landing captures a number of these pannes on the marsh plain. *(a: Rodgers and Kerr 1860, courtesy of NOAA; b: Annex photo 22833, courtesy of Sonoma Heritage Collection)*

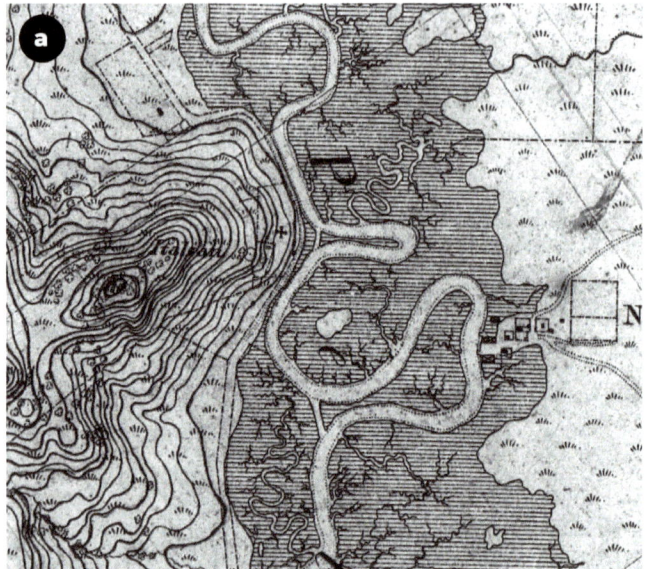

TRANSITION ZONE

The tidal-terrestrial transition zone denotes the area of interaction between tidal and terrestrial or tidal and fluvial processes that influenced the formation of distinctive wetland habitats and supported unique assemblages of plants and animals. The edge of the Petaluma Marsh measured just over 80 km (50 mi); steep (hillslope) interfaces formed approximately half (53%) of the marsh edge, while flatter (alluvial plain) interfaces formed the remainder (Fig. 3.7). Within the flatter interfaces, wet meadow occupied approximately 60% of the marsh perimeter. Adjoining upland habitats, while not included in the historical synthesis mapping and not the focus of research for this study, were likely composed largely of grassland and oak woodland (Thompson 1877a, Barrett 1908).

The Petaluma Marsh historically received direct freshwater inputs from the fluvial Petaluma River to the north, from San Antonio Creek which drained from the west, and from Thompson Creek which connected to the river corridor in the upper estuary. Other creeks dissipated into the extensive wet meadows that bordered many segments the marsh perimeter (Fig. 3.8). As a result of the Petaluma River's relatively small watershed size, the quantity of freshwater input to the Petaluma Marsh was likely low relative to the marshes at the mouth of the Napa River to the east (Vasey et al. 2012).

The interaction between these freshwater sources and the tidewater produced brackish conditions that provided habitat for a diversity of fish and wildlife. Mid-nineteenth century reports indicate that smelt, considered to be a "true estuarine resident that does not fit well into either the euryhaline marine or obligatory freshwater types," were caught in abundance in San Antonio Creek (Parker 1869, Leidy 2007). Similarly, terrestrial-tidal transition zones provided crucial habitat for ecotone specialists that moved between the marsh and upland. Several shrew species[1] inhabited the historical Petaluma Marsh and depended on the middle marsh-high marsh ecotone for nesting and foraging (Samuels

---

[1] Ornate shrew, Sonoma shrew

Figure 3.7. Contemporary LiDAR shows the T-zone gradient between the marsh edge and the surrounding habitats. Just over half of the marsh edge is met by steep upland habitat while the remainder is bordered by low-lying areas, 60% of which is wet meadow. *(County of Marin 2015)*

Figure 3.8. Fluvial-tidal interface types describe the relationship between freshwater inputs and the estuary. Most streams draining towards the estuary historically were disconnected channels that dissipated on the alluvial plain or in freshwater wetlands before reaching the tidal marsh. The remaining streams were either connected directly to the tidal channel network, or drained onto the tidal marshland without connecting to the tidal channel network. Interface data shown here are from SFEI-ASC 2017, and include lower-order channels not shown in the historical synthesis mapping.

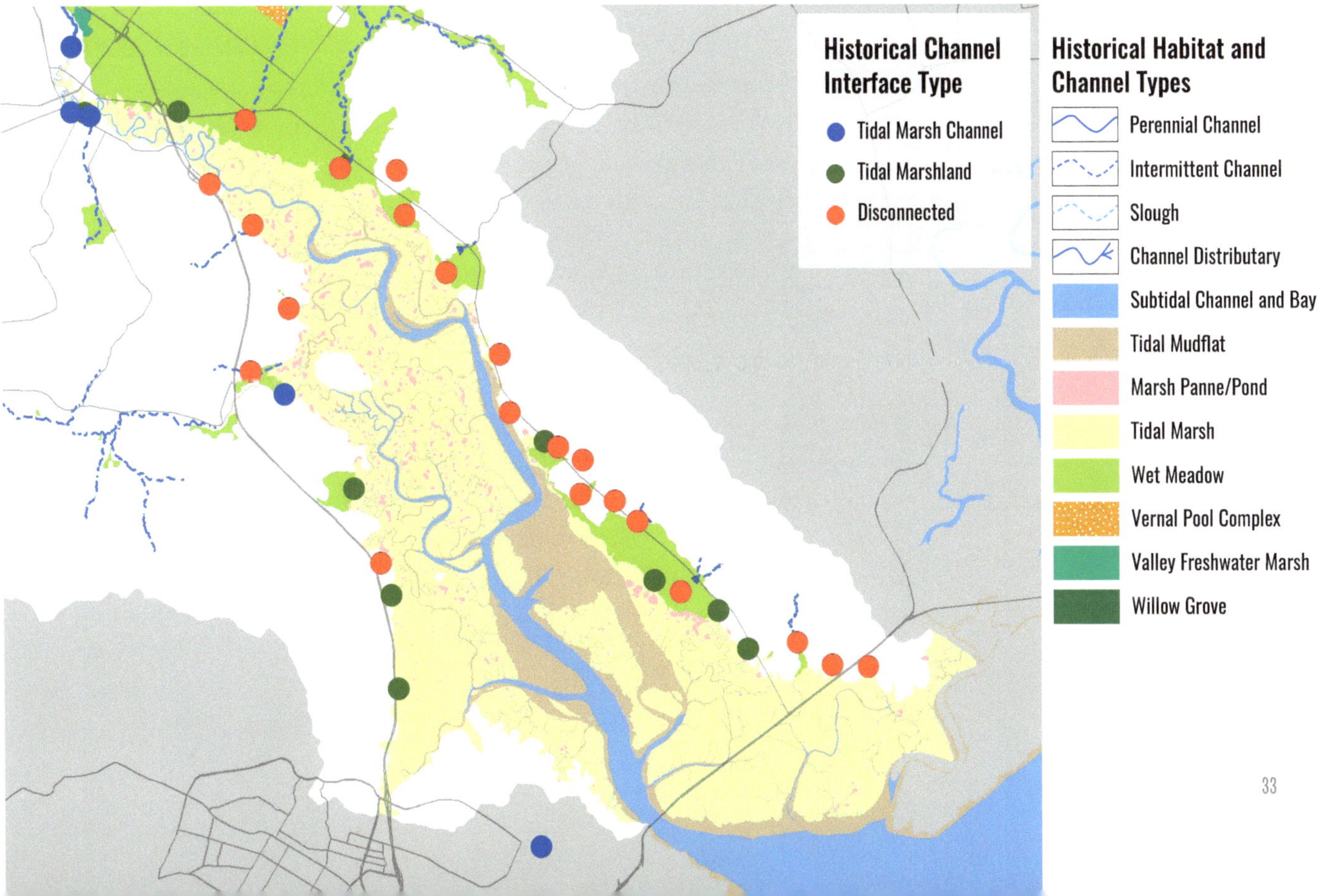

1856d, Feathers 1935, Shellhammer 2012). In December 1908, for instance, zoologist Joseph Dixon noted, "I caught the two *Sorex* [*ornatus californicus*] on the upper edge of the marsh along the railroad track. The marsh at this place is brackish as the water is largely fresh" (Dixon 1908).

FISH AND WILDLIFE

The Petaluma Marsh and its tidal channels provided important habitat for a number of fish and wildlife species that depended on the productive estuarine landscape to live, feed, nest, and reproduce. The salinity gradient and tidal flux in the marsh channels contributed to a diverse fish community, while the mosaic of marsh habitats supported resident and migratory birds and various mammals.

Salmon trout are swarming in Petaluma Creek.
 —Morning Call 1890

Narrative accounts of estuarine fish species within the watershed historically include sturgeon, steelhead, and smelt (Parker 1869 and 1870, *Morning Call* 1890). In 1878, the *Healdsburg Enterprise* (1878) reported sturgeon to be "very plentiful in Petaluma Creek," while in October 1890 the *Morning Call* (1890) stated that "salmon trout [likely steelhead] are swarming in Petaluma Creek." Early records from zoological collections and California Department of Fish and Game surveys also documented native estuarine species such as Sacramento splittail, prickly sculpin, and threadfin shad (Leidy 2007, UCANR 2017).

The productivity of the tidal estuary also made it an attractive habitat for a variety of shorebirds, waterfowl, and other birds. Ridgway's rail, sora, killdeer, Wilson's snipe, yellowlegs, sandpipers (least, spotted, Western), semipalmated plover, northern shoveler, teal, geese, marsh wren, San Pablo song sparrow, white-tailed kite, northern harrier, and short-eared owl were among the birds reported in the marsh historically (Newberry 1857, Parker 1868 and 1869, Dixon 1908, Lamb 1927). Newberry (1857) reported that Ridgway's rails (known as "mud hens") were "very numerous" in Petaluma Marsh. Birds exploited the estuary's tidal regime, utilizing different habitats and elevations throughout the day to forage. Zoologist Joseph Dixon, for instance, observed that San Pablo song sparrows—a tidal marsh obligate endemic to San Pablo Bay that is currently listed as a species of special concern in California—"scatter out in the sloughs at low tide and it is very hard to find them, but at high tide they come out on the hard land" (Dixon 1908). During the fall and winter, large numbers of northern pintail, American wigeon, canvasback, and other waterfowl migrated to the marsh to overwinter and form nesting pairs (Lee 1901, National Audubon Society 2017).

The marsh also provided abundant and varied habitat for a number of mammals. Nineteenth and early twentieth century collectors noted shrew and California vole moving between the pickleweed and the upper marsh grass, as well as salt marsh harvest mouse, which was observed extensively throughout the marsh plain from Burdell Island to McNear's Bridge (Dixon 1908; Fig. 3.9). California lowland mink and northern raccoon were also observed using marsh habitats (Dixon 1908).

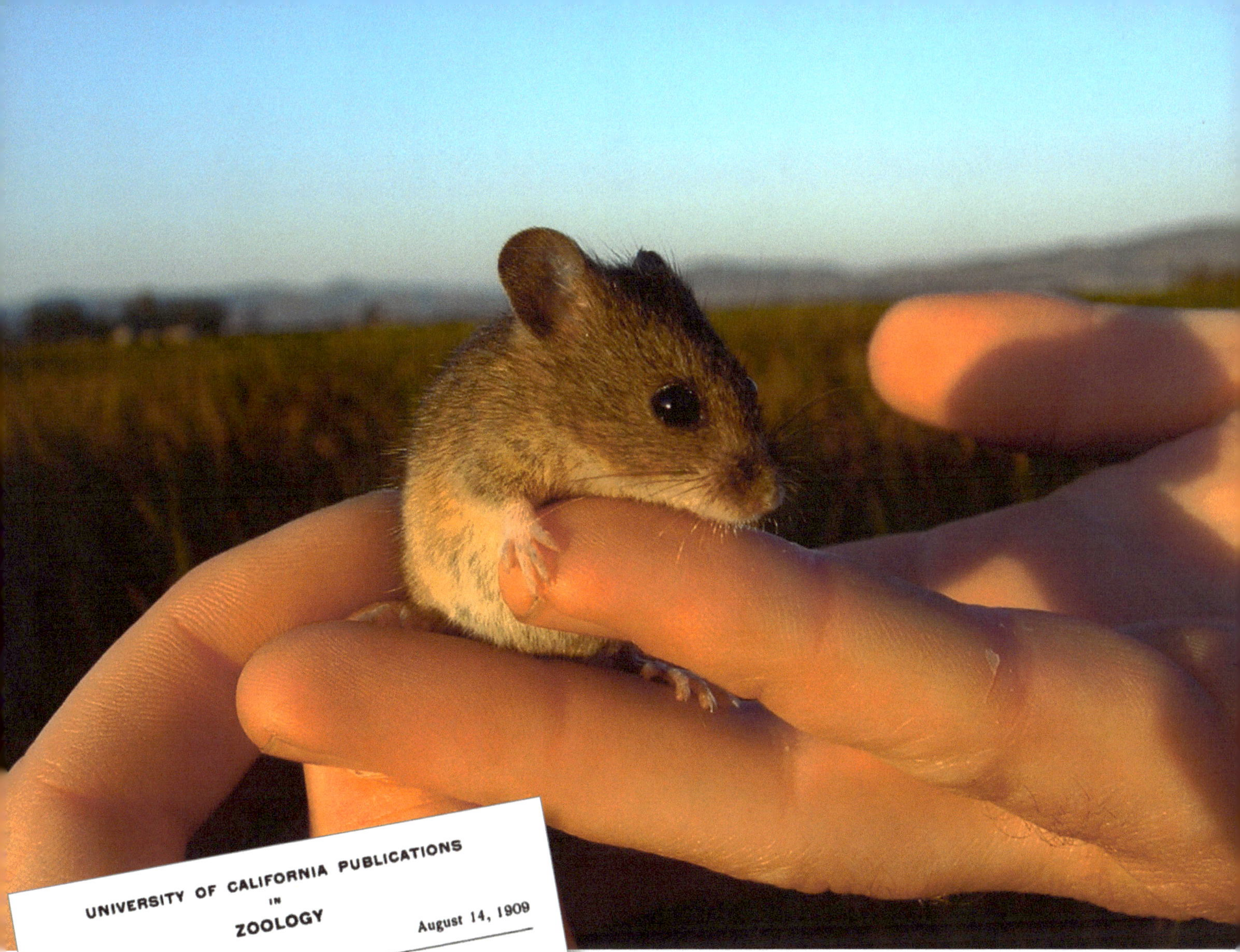

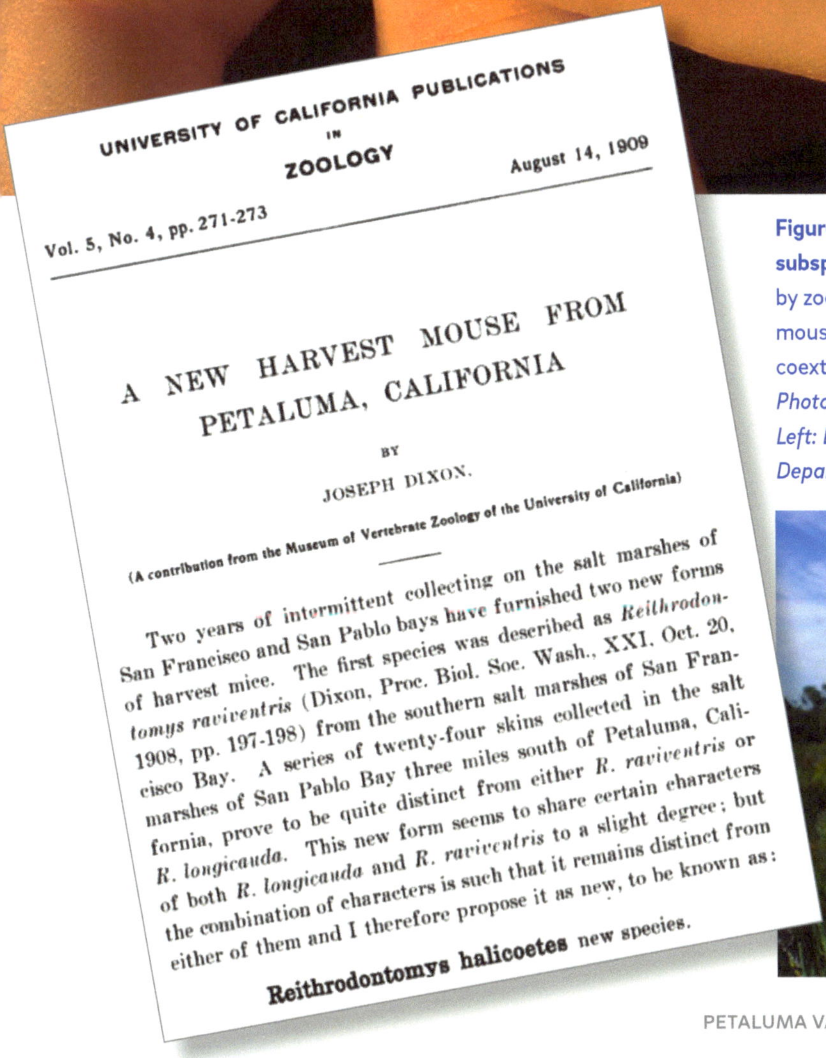

UNIVERSITY OF CALIFORNIA PUBLICATIONS
IN
ZOOLOGY

Vol. 5, No. 4, pp. 271-273 August 14, 1909

A NEW HARVEST MOUSE FROM PETALUMA, CALIFORNIA

BY

JOSEPH DIXON.

(A contribution from the Museum of Vertebrate Zoology of the University of California)

Two years of intermittent collecting on the salt marshes of San Francisco and San Pablo bays have furnished two new forms of harvest mice. The first species was described as *Reithrodontomys raviventris* (Dixon, Proc. Biol. Soc. Wash., XXI. Oct. 20, 1908, pp. 197-198) from the southern salt marshes of San Francisco Bay. A series of twenty-four skins collected in the salt marshes of San Pablo Bay three miles south of Petaluma, California, prove to be quite distinct from either *R. raviventris* or *R. longicauda*. This new form seems to share certain characters of both *R. longicauda* and *R. raviventris* to a slight degree; but the combination of characters is such that it remains distinct from either of them and I therefore propose it as new, to be known as:

Reithrodontomys halicoetes new species.

Figure 3.9. Petaluma Marsh is the type locality for the northern subspecies of salt marsh harvest mouse, which was described by zoologist Joseph Dixon in 1909. Dixon (1909) states, "This mouse seems to be restricted to the salt marsh, its range being coextensive with that of the 'pickle grass' (*Salicornia*)." *(Above: Photo by USGS, May 2004, licensed under Creative Commons; Left: Dixon 1909, courtesy of Google Books; Below: Photo by Department of the Interior, courtesy of Wikipedia)*

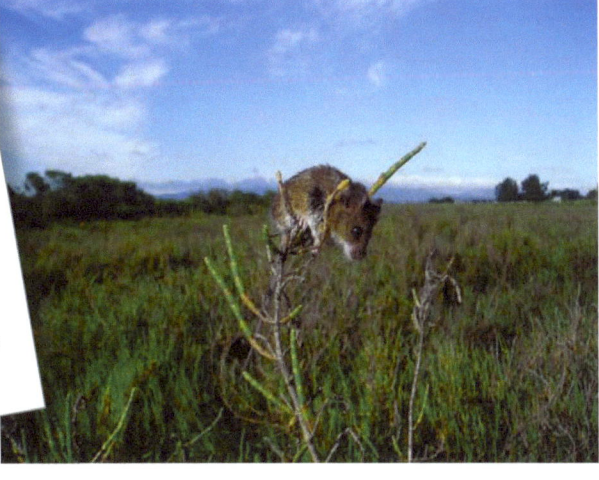

Petaluma River. *(photo by Sean Baumgarten, July 20, 2017)*

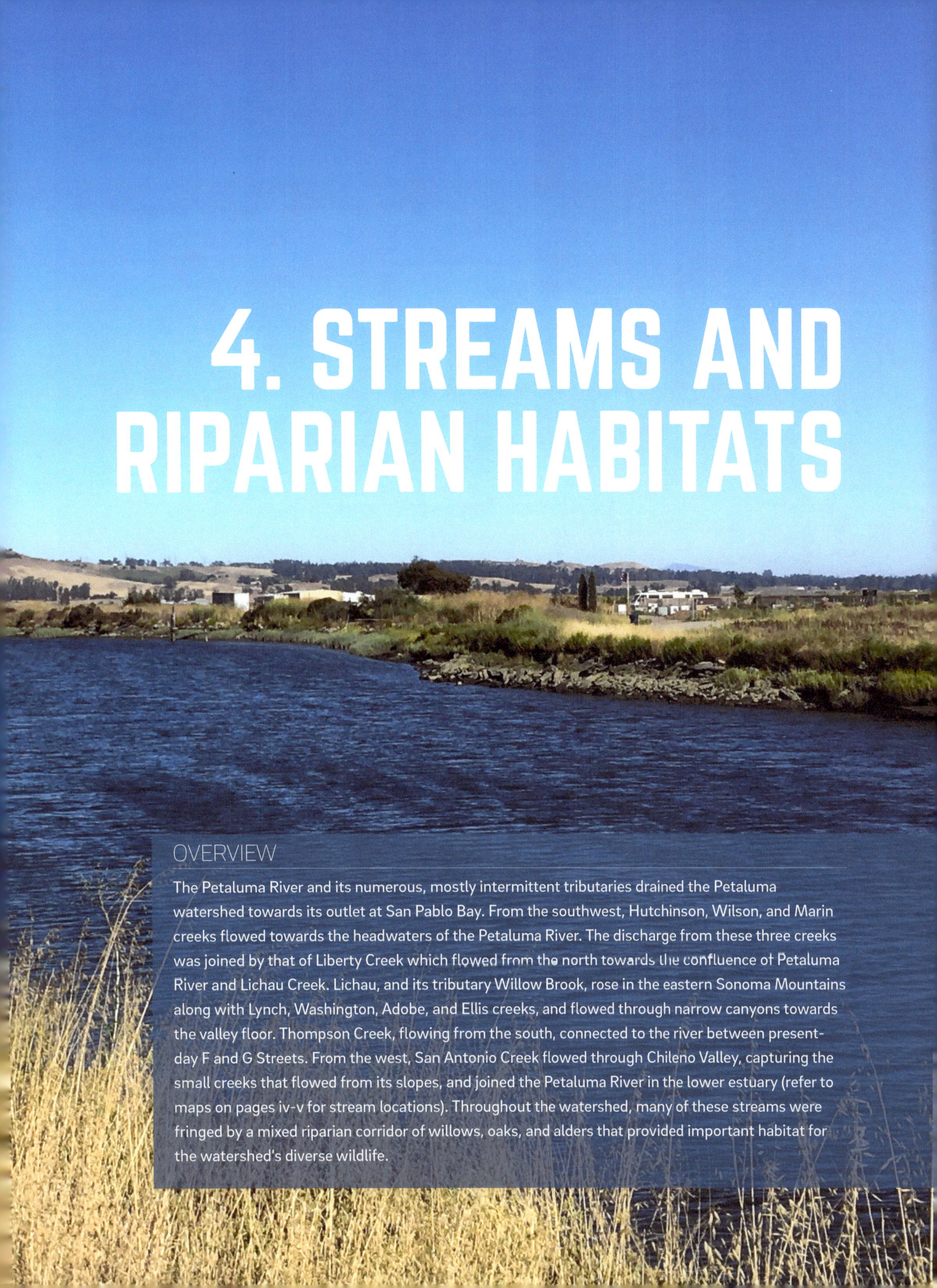

4. STREAMS AND RIPARIAN HABITATS

OVERVIEW

The Petaluma River and its numerous, mostly intermittent tributaries drained the Petaluma watershed towards its outlet at San Pablo Bay. From the southwest, Hutchinson, Wilson, and Marin creeks flowed towards the headwaters of the Petaluma River. The discharge from these three creeks was joined by that of Liberty Creek which flowed from the north towards the confluence of Petaluma River and Lichau Creek. Lichau, and its tributary Willow Brook, rose in the eastern Sonoma Mountains along with Lynch, Washington, Adobe, and Ellis creeks, and flowed through narrow canyons towards the valley floor. Thompson Creek, flowing from the south, connected to the river between present-day F and G Streets. From the west, San Antonio Creek flowed through Chileno Valley, capturing the small creeks that flowed from its slopes, and joined the Petaluma River in the lower estuary (refer to maps on pages iv-v for stream locations). Throughout the watershed, many of these streams were fringed by a mixed riparian corridor of willows, oaks, and alders that provided important habitat for the watershed's diverse wildlife.

Figure 4.1. The Petaluma River maintained a relatively straight, single-threaded channel for most of its length upstream of the estuary. (map at 1:40,000) *(NAIP 2016)*

The fluvial portion of the Petaluma River extended approximately 6 km (3 mi) from its confluence with Lichau Creek downstream to the inland edge of the estuary (Fig. 4.1). Confined by bedrock on the west and bordered by extensive wet meadow on the east (see pages 56–59), the river maintained a relatively straight course towards the southeast over most of its length. Near the present-day confluence with Lynch Creek, the river turned towards the south and flowed through the City of Petaluma, where it entered the estuary. The mainstem was characterized by a single-threaded channel punctuated by numerous large in-channel pools that may have provided cold-water refugia for native fish. Floods were common along the mainstem during the winter, but during the dry season flows were likely quite low, and some reaches may have experienced an intermittent flow regime.

The largest tributary to the Petaluma River was, and still is, San Antonio Creek, which drained approximately 95 km² (37 mi²) on the southwestern side of the watershed and emptied into the estuary via San Antonio Slough. The creek originated at Antonio Mountain and flowed into the Laguna de San Antonio, a

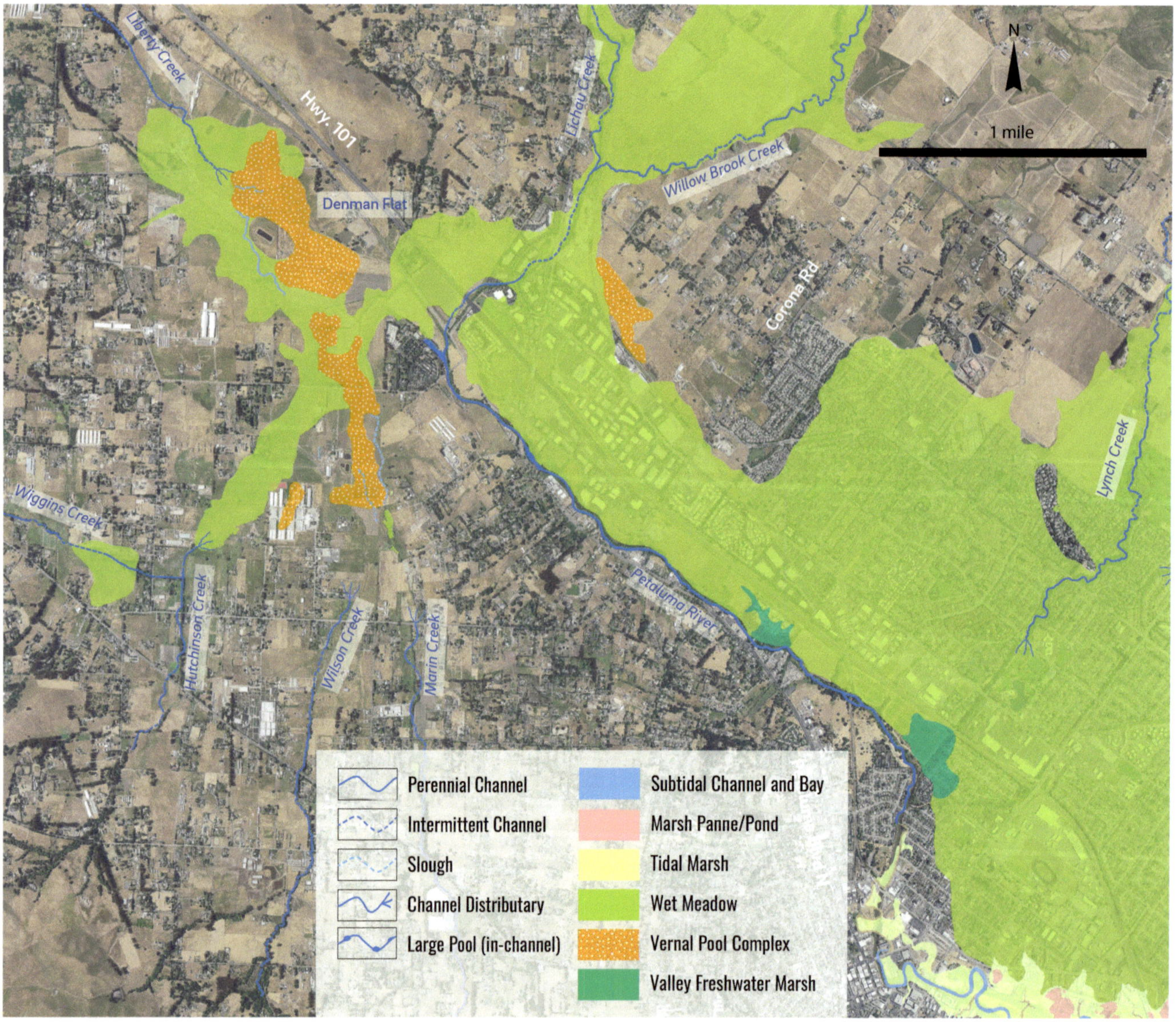

38 CHAPTER 4 • STREAMS AND RIPARIAN HABITATS

wetland complex near the drainage divide with Walker Creek (see pages 64–66). From the Laguna at the top of the San Antonio subwatershed, the creek flowed through Chileno Valley and captured a number of small tributaries along its course. For most of its length, San Antonio Creek ran through a bedrock canyon ranging in width from about 30–1200 m (92–3871 ft). Portions of the channel were likely dry during the summer months, though certain reaches may have sustained limited perennial streamflow.

This chapter describes hydrologic patterns within the watershed, focusing on flood dynamics, variability in streamflow, and hydrologic connectivity, and summarizes historical information pertaining to riparian and aquatic habitats.

FLOODING AND FLOW VARIABILITY

The Mediterranean climate of the Petaluma River watershed, characterized by mild winters and dry summers, resulted in pronounced seasonal variability in streamflow. During the wet season, flooding was a common occurrence in low-lying areas throughout the watershed, such as along the Petaluma River mainstem, on the alluvial plain east of the river, and in the wetland complex at the downstream ends of Liberty, Hutchinson, Wilson, and Marin creeks (Figs. 4.2, 4.3, 4.4; Heuer 1917). Denman Flat, located next to present-day Redwood Highway South and Stony Point Road, appears frequently in the chronicles of flood events as one of the most vulnerable zones in the floodplain. An 1896 report compiled information on maximum flood discharges from creeks draining towards the City of Petaluma, and concluded that maximum discharge typically did not exceed 3,000 cfs for Petaluma Creek, 650 cfs for Washington Creek, and 1,000 cfs for Lynch Creek (Price and Nurse 1896).

The magnitude of early flood events are often described in the historical record in terms of their impact on the growing City of Petaluma. Reporting on a flood in December of 1871, for example, the *Sonoma Democrat* stated, "At Petaluma both sides of the creek are overflowed and a perfect torrent is rushing down Washington Street. The water reaches the lower floors of the warehouses and buildings in East Petaluma" (*Sonoma Democrat* 1871). When the magnitude of flooding was at its height, flood waters in low-lying areas like Denman Flat and the Cotati Rancho could extend laterally for up to several kilometers (*Petaluma Weekly Argus* 1881a, *Healdsburg Tribune* 1925). In December 1881 the *Petaluma Weekly Argus* reported:

> The recent rain storm, which began Friday night last, and continued almost unabated till Monday morning, was the heaviest ever experienced in the State. The water rose until almost all of the eastern and southern part of town was submerged. The water was found by our gauge to be about twelve inches higher than ever before known…In the lower part of town travel was only possible by the aid of row boats. (*Petaluma Weekly Argus* 1881b)

Early accounts describe repeated flooding in Eastern Petaluma and the area upstream of the Washington Street Bridge, which was determined to be "inadequate for conducting floodwaters" (Price and Nurse 1896). During heavy

Coming as they do from an abrupt, hilly watershed, they rush from the cañons [sic] that confined them, with torrential velocity, to a gentle sloping watershed, where they must necessarily spread beyond natural confines in flood-periods.

—Price and Nurse 1896, describing flooding along Reservoir, Edwards, and Thompson creeks

Figure 4.2. (above) This photo was taken during a 1904 flood from the present-day site of the Petaluma Yacht Club looking upstream towards Water Street. The original brick flour mill at 148 Petaluma Boulevard North and the bank flag pole can be seen in the background and are still visible today. *(Photo ID # 1995-08-02, Negative 23-20, courtesy of Petaluma Historical Library & Museum)*

Figure 4.3. (left) A 1912 flood event inundates Camm & Hedges Co., Inc. at 960 Washington Street, half a kilometer (0.3 mi) from the Petaluma River. A horse and buggy tows people through the submerged streets while a man in the background rows a boat through the floodwaters. *(Photo ID# 2005-273-18, Flood File 23-20, courtesy of Petaluma Historical Library & Museum)*

During the recent high water the large brick warehouse of John A. McNear, in East Petaluma, stood like an oasis in the desert of waters.

—*Petaluma Weekly Argus* 1881a

Figure 4.4. (above) This 1912 photo looks east from the Washington Street Bridge during a large flood event. A horse and buggy wades through the floodwaters with the Golden Eagle Mill in the background. *(Photo ID # 1995-80-01, Negative 23-20, courtesy of Petaluma Historical Library & Museum)*

floods, water could reach a depth of 0.8 m (2.5 ft) on the bridge (Price and Nurse 1896), and perhaps even higher in adjacent areas: flood waters on Washington Street during a 1904 flood were reportedly higher than the wheels of a buggy (*Marin Journal* 1915). Resident John R. Stone recalled how flood waters in the 1920s or 30s inundated the Golden Eagle Mill just south of Washingston Street, washed away all the lumber in Gold's landing, and rose up to the bellies of horses; during large storms, flood waters could take several days to recede (Stone and Curme 1975; Fig. 4.4). Floods delivered large volumes of water and sediment to the estuary, and were part of the early motivation for channel widening and other modifications (see pages 76–79).

During the dry season, streamflow within alluvial portions of the watershed was limited, and many creeks ceased to flow entirely. Limited perennial flow or standing water likely occurred along portions of the Petaluma River mainstem, particularly in reaches that supported large in-channel pools (see pages 52–53) or were adjacent to patches of valley freshwater marsh (see pages 64–66; USGS [1914]1916). Extensive seasonal wetlands, which occupied much of the alluvial basin upstream of the river and the alluvial plain east of the river, may have provided some surface or subsurface input to portions of the river throughout much of the year, though the volume of dry season flow was generally quite small (Adams et al. 1912, Heuer 1917). San Antonio Creek may have also supported perennial flow in some reaches (Collins et al. 2000). Though smaller tributaries to San Antonio Creek were dry during the summer (Thompson 1864), the Laguna wetland complex at the head of the drainage would have provided baseflow to the

The fresh-water portions of Petaluma Creek and its other tributaries have very little flow during the dry season.

—Heuer 1917

We descended to the plain, and presently came to the stream, which, by the Indians and men of our company who had seen the same on several occasions, is considered as the most copious of all in this locality, and we found it to be without water and entirely dried up in coming on the plain; although at the foot of the hills, where it runs down, there was a little rill, but so small as to be altogether unpromising.

—Altimira 1823, describing a tributary of the Petaluma River

creek during the dry season, and may have maintained surface flow in some reaches year-round.

The lower portions of creeks draining the Sonoma Mountains on the eastern side of the watershed—Lichau, Willow Brook, Lynch, Washington, Adobe and Ellis creeks—were generally dry during the summer months. In late June 1823, for example, Spanish explorers in the Altimira expedition lamented the absence of water in the valley, describing their disappointment in discovering that a stream, supposed to be "the most copious of all," had dried up completely (Altimira 1823). Though streamflow on the alluvial plain was intermittent, springs maintained perennial flows in the upstream reaches of several creeks (Fig. 4.5; Lynch et al. 1872). The upper reaches of Adobe, Lynch, and Copeland creeks, for instance, which served as water sources for the City of Petaluma in the late 1800s, were reported to have ample dry season flow to support water supply needs (*Petaluma Weekly Argus* 1877a, *Petaluma Weekly Argus* 1877b, Irelan 1893).

Figure 4.5. Several tributaries, including Lichau Creek (below, left) and Adobe Creek (below, right), maintained perennial flow in upstream reaches (depicted by a solid line in the early USGS quad), but transitioned to intermittent flow as they reached the valley floor (depicted by a dashed line). During the summer months, these creeks would have run dry as they approached the valley floor, but in times of heavy rainfall they swelled with floodwater and conveyed flows towards the estuary downstream. *(USGS [1914]1916)*

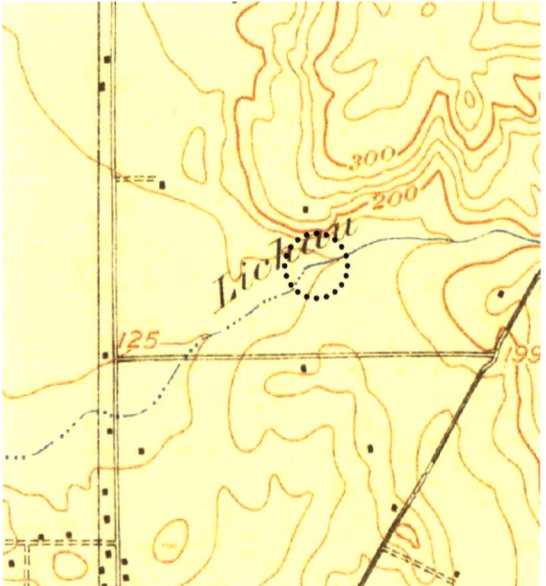 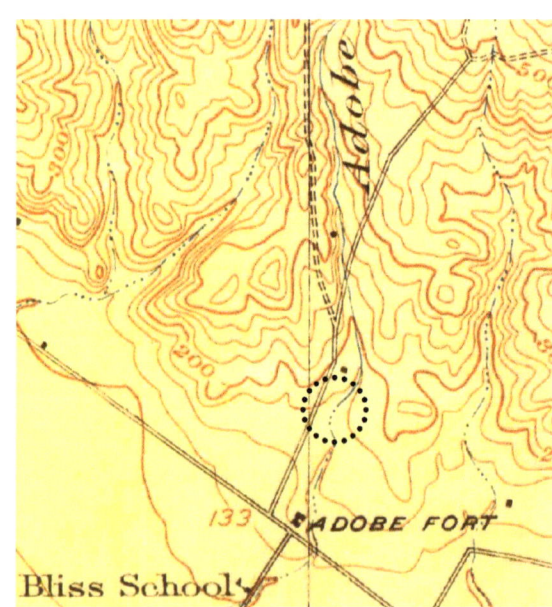

HYDROLOGIC CONNECTIVITY

Historically, few streams within the watershed maintained defined channels directly into the Petaluma River mainstem or its estuary. Like many systems with low stream power around the Bay, smaller tributaries within the watershed dissipated on alluvial plains upstream of the estuary, often spreading into multiple distributary channels (SFEI-ASC 2017). For example, most of the streams that originated in the Sonoma Mountains on the eastern side of the Petaluma River, including Lynch, Washington, East Washington, Adobe, and Ellis creeks, lost definition as they reached the broad alluvial plain that bordered the mainstem corridor and much of the estuary (Fig. 4.6). To the northwest of the city, Hutchinson, Wilson, Marin, and Wiggins creeks flowed northward and dispersed into a wet meadow-vernal pool complex in the Denman Flat area between present-day Skillman Lane and Rainsville Road (see pages 57-58). Liberty Creek, which drained the northern part of the watershed west of Meacham Hill, also dissipated within this wetland complex (Fig. 4.7). These discontinuous tributaries provided freshwater and sediment input that would have helped sustain seasonal wetlands at their downstream ends. In many cases, shallow emergent sloughs transported flows through the wetland complexes downstream of channel distributaries (Fig. 4.8).

Unlike many of the smaller tributaries in the watershed, creeks with larger flows and higher stream power maintained defined channels to their confluence with the Petaluma River or the tidal wetlands.

Figure 4.6. Most of the creeks draining the Sonoma Mountains lost definition before reaching the Petaluma River and estuary and dissipated in an extensive wet meadow that occupied the alluvial plain on the eastern side of the valley. This 1866 county map shows Lynch ("Alder"), Washington, and Adobe creeks terminating upstream of the estuary. *(Bowers 1866, courtesy of David Rumsey Map Collection)*

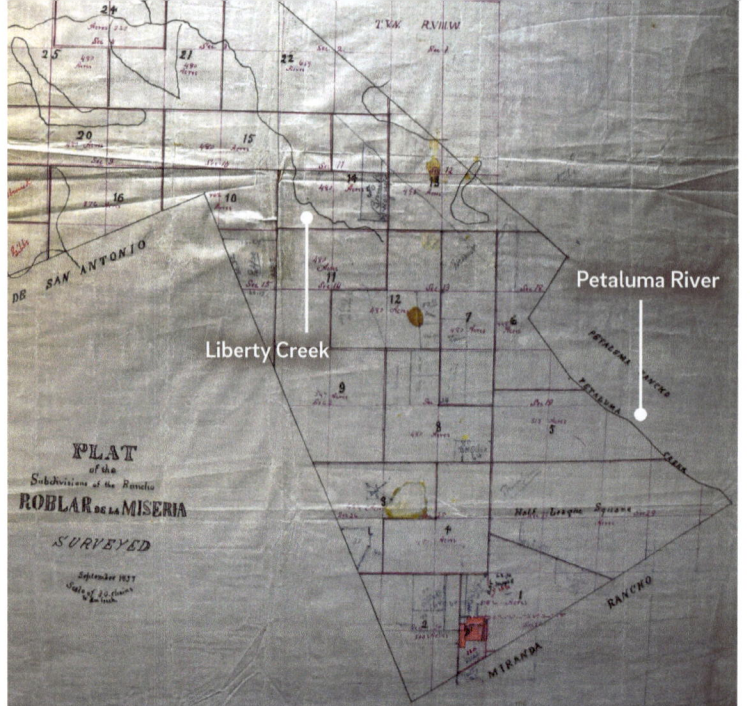

Figure 4.7. As depicted in this 1857 map of Rancho Roblar de la Miseria, Liberty Creek was generally disconnected from the Petaluma mainstem and discharged into a wetland complex near the headwaters of the Petaluma River. *(Unknown 1857, courtesy of Curtis & Associates, Inc.)*

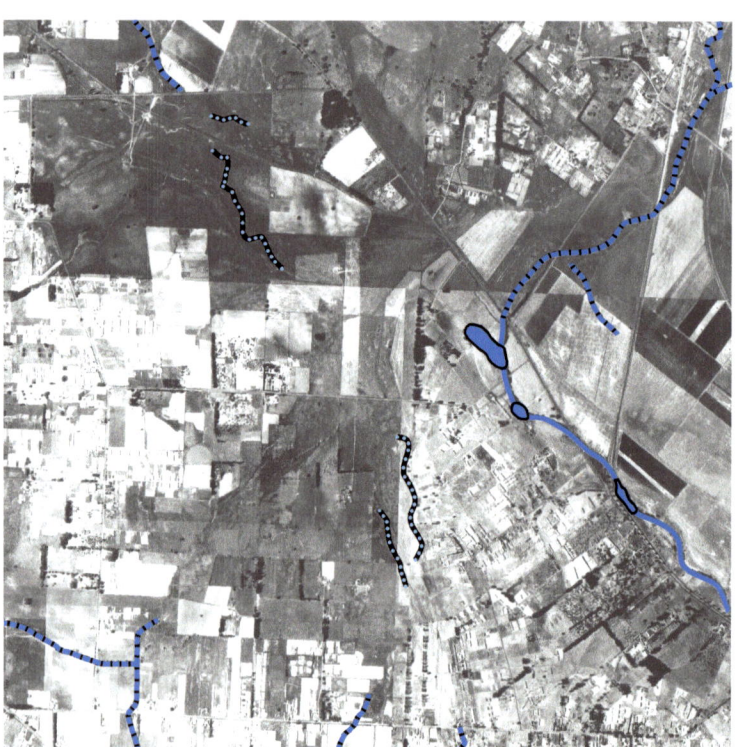

Figure 4.8. (left) Several small emergent sloughs are visible in the wetland complex near the Petaluma River headwaters in the 1942 aerial photos. (Map at 1:8000) *(USDA 1942)*

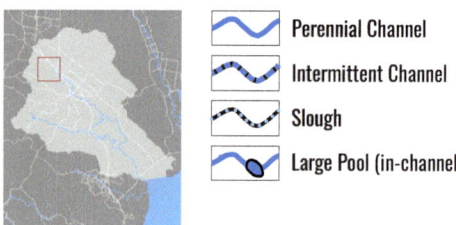

Figure 4.9. (below) Unlike many of the smaller tributaries in the watershed, San Antonio Creek maintained a defined channel that connected directly to the estuary. An 1859 plat of Rancho Olompali shows the connectivity between the fluvial portion of San Antonio Creek (labeled "Arroyo San Antonio") and San Antonio Slough within the tidal marsh (labeled "Estero San Antonio"). *(Matthewson 1859b, courtesy of Bureau of Land Management)*

San Antonio Creek, for instance, connected with the tidal wetlands at San Antonio Slough, which wound its way through the marsh for approximately 11 km (7 mi) and joined the tidal Petaluma River south of Neil's Island (Fig. 4.9). Upstream of the estuary, Lichau Creek was the principal tributary of the Petaluma River, and the only major stream that connected directly to the mainstem historically (Cardwell 1958). At times of high rainfall, Copeland Creek, which drains from Sonoma Mountain and typically flowed north toward the Laguna de Santa Rosa, overflowed into the Petaluma River watershed and joined with Lichau Creek (Fig. 4.10; Bowers 1866, Price and Nurse 1896, Holway 1907).[1] According to Holway (1907):

> Copeland Creek flows down the western slope of Sonoma Mountain and debouches on a fan that spreads out over the flat divide separating the Russian River from the Bay. The southernmost of the distributaries on this fan empties into Petaluma Creek and thence to the bay. The northernmost flows into the Russian River. These distributaries meet today at the head of the fan and in flood time Copeland Creek discharges both ways.

Despite having only an episodic connection to the watershed, Copeland Creek may have been an important source of sediment input to the Petaluma River: a 1917 Army Corps report reported that "it is claimed by local interests that the greater part of this gravel, sand, and silt is brought into Petaluma Creek by Copeland Creek...in times of flood" (Heuer 1917).

A hydrologic surface connection also existed between San Antonio Creek and the Walker Creek watershed to the west. From its headwaters at Antonio Mountain, the uppermost reach of San Antonio Creek flowed northwards into the Laguna de San Antonio (see pages 64–66) near the boundary with

Figure 4.10. (below) Copeland Creek is depicted with a forked course that crosses the drainage divide on this 1866 county map. *(Bowers 1866, courtesy of David Rumsey Map Collection)*

[1] Copeland Creek overflowed into the Petaluma River Watershed as recently as 2005 (Dawson and Sloop 2010).

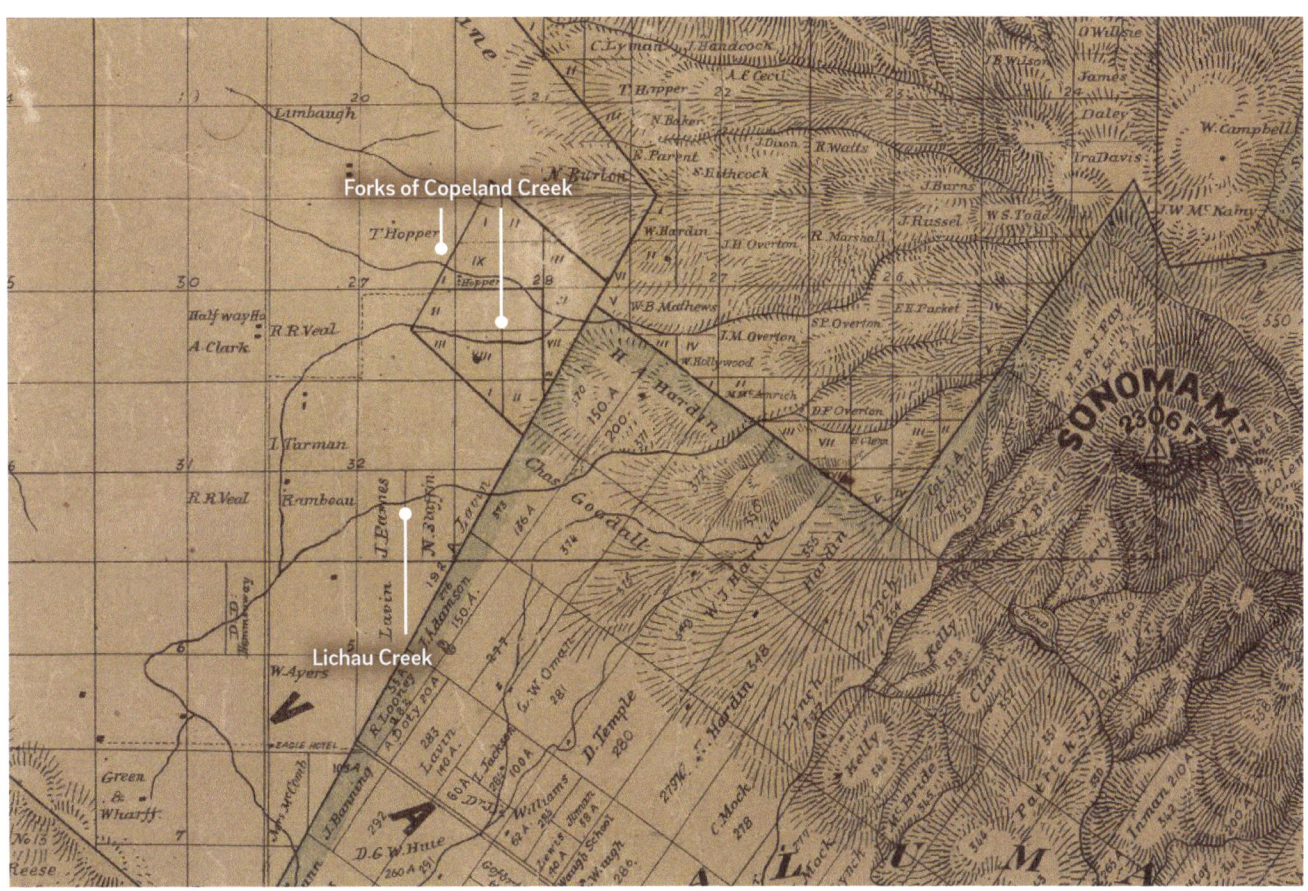

the adjacent Walker Creek watershed. The relatively flat topography in this area made for a very subtle drainage divide which, during flood events, established connectivity between San Antonio Creek, Laguna Lake (in the Walker Creek watershed), and the Laguna de San Antonio (Fig. 4.11). Holway (1914) stated that this drainage divide "is so nearly level that according to the reports of residents the rivulets of the rainy season flow toward the bay or toward the ocean, according to the irregularities of the last plowing" (Fig. 4.12). Geologic evidence indicates that the upper reaches of the San Antonio Creek watershed may have formerly comprised the head of the Walker Creek drainage, and were "captured" by San Antonio Creek following a period of uplift during the late Pleistocene (Holway 1914, Dickerson 1922).

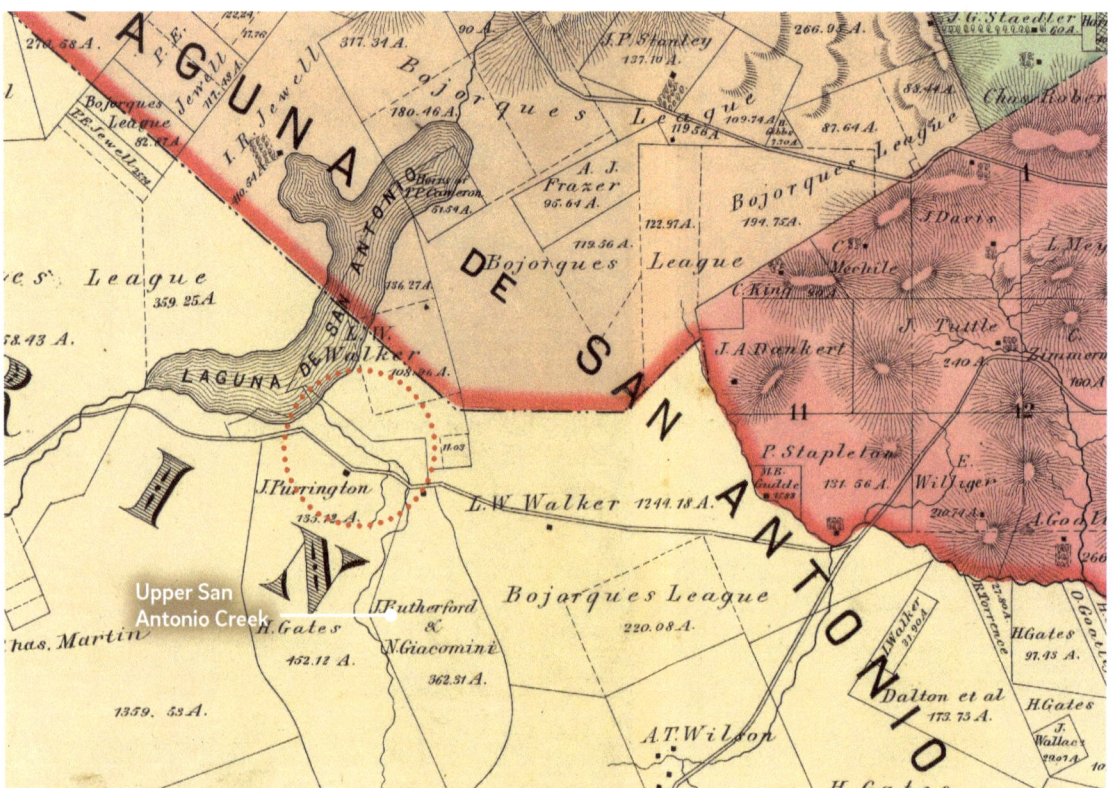

Figure 4.11. left) **The flat topography near the Laguna de San Antonio occasionally resulted in a hydrologic connection** between San Antonio Creek and the Walker Creek watershed to the west. This 1877 county map shows the upper reach of San Antonio Creek above the Laguna flowing west across the drainage divide. *(Thompson 1877b, courtesy of David Rumsey Map Collection)*

Figure 4.12. (right) **This ca. 1914 image, looking westward across the drainage divide at the head of San Antonio Creek,** shows the extremely flat topography of the area. Upper San Antonio Creek, ditched by this time, is visible in the foreground, and Laguna Lake (in the Walker Creek watershed) is visible at the base of the hills in the background. *(Holway 1914, Plate 15 [top], courtesy of University of California Publications in Geography)*

RIPARIAN AND AQUATIC HABITATS

Though riparian corridors were not mapped as part of this study, the collected data provide insights into riparian vegetation composition in many parts of the watershed. Nineteenth and early twentieth sources indicate that many creeks supported a mixed riparian forest dominated by willow, alder, oak, and California bay laurel. These riparian corridors provided important ecological functions, such as wildlife habitat, bank stability, channel shading, and large woody debris inputs.

Along San Antonio Creek, GLO surveyors recorded evidence of a mixed riparian corridor dominated by oaks, laurels, willows, and buckeyes along the entire length of the creek and its tributaries (Fig. 4.13; Matthewson 1859a,b). These early records are consistent with early photographs of the creek (Fig. 4.14) and with the vegetation type mapping (VTM) performed by Albert Everett Wieslander and the U.S. Forest Service in the 1930s, which lists willow, oak, and California bay laurel as the dominant species along downstream reaches of San Antonio Creek (Wieslander 1930). Abundant anecdotal evidence also supports this interpretation: Gelo Parker described willows and laurels along San Antonio Creek in the 1860s (see page 49), while in the early 20th century, zoologist Joseph Dixon recounted his day setting traps along the creek, noting that the foothill canyons were "well wooded with laurel, white and live oak" (Dixon 1908).

In comparison with San Antonio Creek, relatively little historical data is available with regard to riparian habitats along the Petaluma River or other tributaries. Mid-nineteenth century land grant maps indicate the presence of riparian forests along much of the Petaluma River mainstem; portions of Willow Brook, Lynch, Washington, and Adobe creeks; and tributaries of San Antonio creek (Fig. 4.15). Sporadic early evidence suggests that willows, oaks, and alders were among the trees comprising these riparian corridors (USDC 1852b, *Sonoma County Journal* 1855, Thompson 1857b, Martin 1862, Bowers 1866). Late 19th and early 20th century botanical records

These hills are quite barren and denuded of timber. In the hollows and ravines, there is considerable oak and a few other trees generally fit only for firewood, and dotted along the tops a species of stunted live-oak, and another of evergreen—a kind of laurel—which gives a pleasing and picturesque view to their appearance.

—*California Star* 1848, describing the mountain range separating the Petaluma and Sonoma valleys

Figure 4.13. Oaks, California bay laurels, willows, and buckeyes were among the dominant riparian species observed along the San Antonio Creek in the 19th and early 20th centuries. These corridors provided important ecosystem functions including bank stability, riparian wildlife habitat, and shading. *(Matthewson 1859b, courtesy of Bureau of Land Management)*

Figure 4.14. (left) **A narrow mixed riparian forest is visible bordering San Antonio Creek** in this view from ca. 1914, which is looking downstream approximately 3–5 km (2–3 mi) below the top of the watershed. *(Holway 1914, Plate 17 [top], courtesy of University of California Publications in Geography)*

Figure 4.15. (below) **Mid-19th century land grant maps suggest the presence of riparian forests along many creeks throughout the watershed.** (a) This 1852 map of Rancho Petaluma shows trees bordering the mainstem Petaluma River, as well as Willow Brook, Lynch, Washington, and Adobe creeks. (b) An 1845 map of Rancho Roblar de la Miseria shows large trees lining the Petaluma River upstream of the estuary. (c) Another map of Roblar de la Miseria from 1852 shows an "oak tree" and "willows" along Willow Brook Creek upstream of its confluence with the Petaluma River. *(a: USDC 1852a, courtesy of The Bancroft Library; b: USDC 1845, courtesy of The Bancroft Library; c: USDC 1852b, courtesy of The Bancroft Library)*

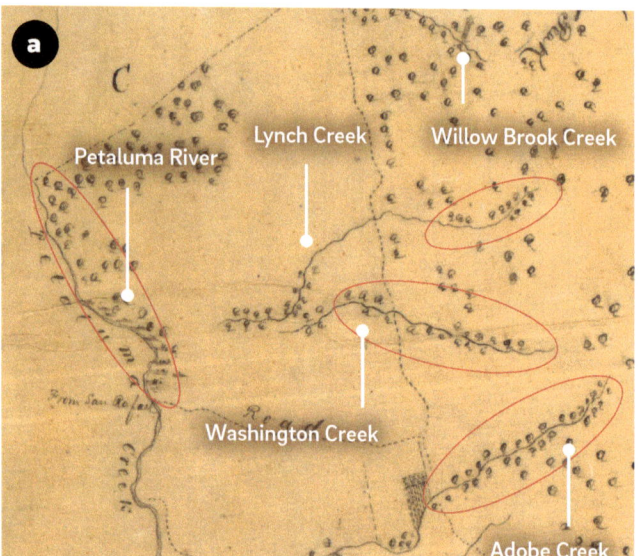

48 CHAPTER 4 • STREAMS AND RIPARIAN HABITATS

The Diary of Gelo Freeman Parker:

One of the most intriguing historical documents of the early Petaluma landscape was a diary that belonged to a young boy named Gelo Freeman Parker. Between 1868 and 1884, Parker wrote daily about his chores on his family's dairy farm 5 km (3 mi) south of the City of Petaluma, as well as his hunting adventures on San Antonio Creek with his older brother, Pitman Wilder Parker. The younger Parker's entries are full of descriptions of riparian trees (e.g., willows and laurels), understory species (e.g., blackberries, elderberries, and gooseberries), and fish and wildlife species that he observed or hunted:

> 1868, September 13: "Set two Quail traps and then go and hunt some. Samuel Heald goes with me. He killed won [sic] quail and I killed two both at [one] shot in a live oak tree."

> 1868, December 26: "Raining the tide came up very high today… Pitman kills a teal."

> 1869, January 10: "Off with Pitman too hunt on the San Antonio we go up the creek oposit the willows and hunt around their on the hils [sic]."

> 1869, February 7: "Take the rifle and go over across the creak [sic] to try and get a shot at some gese [sic] but I cood [sic] not get within 2 or 3 hundred yards of them and they all flew away. I had a bad time getting threw [sic] the mud which had washed down from the hils [sic] and was composed of sand and adobe… It was ankle deap [sic] a good deal of the way up to my neas [sic] and over."

> 1869, February 14: "We go up the San Antonio Creak [sic] we saw 12 hare… In wone [sic] place I saw a squirel [sic] run up a tree and I also saw another threw [sic] the limbs of the lorel [sic] tree."

> 1869, March 12: "I go fishing down to the San Antonio I catch about [8?] small fish and 5 suckers nearly a foot and a half long and the heavist [sic] wone [sic] waid [sic] 2 lb [2?] oz."

> 1870, January 30: "Do my milking and take the shotgun and go across the creek for gras[s] there was plenty there but I could not get any without something to drive them. I kill a mal[l]ard and dress it."

(Parker 1868-1870)

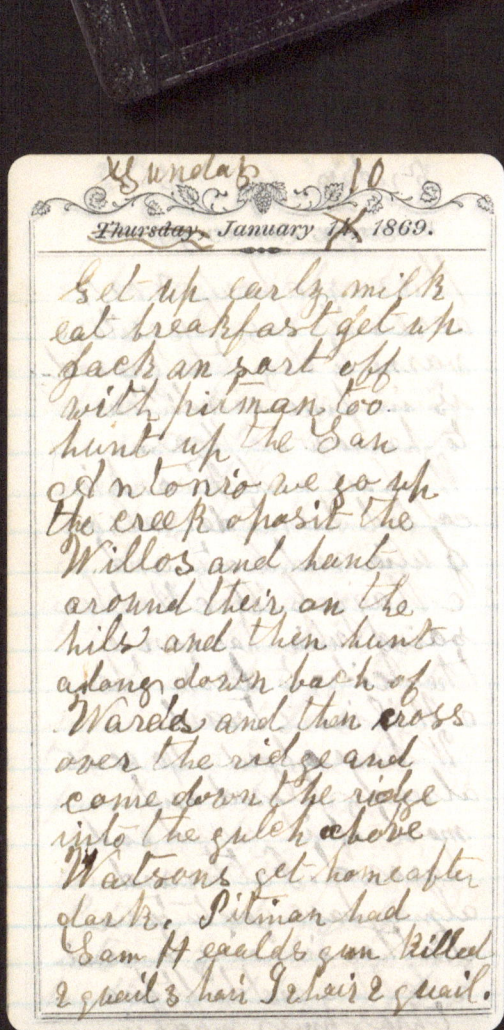

Figure 4.16. A 15-year-old Gelo recounts his adventures along San Antonio Creek with his older brother, Pitman. The brothers hunted the local wildlife and explored the riparian forests along the creek. Gelo's diaries document, in his characteristic phonetic spelling, the game he successfully shot and the trees and undergrowth that he passed through on his way to and from the family home north on the Petaluma Marsh.
(BANC MSS 72/168 c, courtesy of the Bancroft Library)

also document many riparian trees and understory species within the watershed, such as Oregon ash, narrowleaf willow, red willow, mule fat, honeysuckle, and California rose (Table 4.1). The lower portions of Adobe Creek appear to have been characterized by a sparsely vegetated braided channel (Fig. 4.17), which may have also been the dominant channel morphology of other discontinuous tributaries on the alluvial plain.

Riparian forests provided habitat for a diversity of birds, mammals, amphibians, and other wildlife. Early observers made note of numerous bird species known to inhabit riparian forests and streamside thickets, such as "Traill's" (willow) flycatcher, yellow warbler, Lincoln's sparrow, Nuttall's woodpecker, Bullock's oriole, and Lazuli bunting (Samuels 1856a,b,c; Pemberton 1902; Dixon 1908; Lamb 1927). Both foothill yellow-legged frog and the federally threatened California red-legged frog were observed in the northeastern watershed during the early 20th century, and likely occupied streamside habitats (Rodgers 1940, Rodgers and Stirton 1940).

A diversity of fish inhabited the creeks throughout the Petaluma River watershed. Early records document trout, Sacramento suckers, three-spine stickleback, riffle sculpin, and splittails in the mainstem and tributary streams (*Sacramento Daily Union* 1862, Parker 1869, Stone and Curme 1975, Leidy 2007, UCANR 2017). Many of these species are known to inhabit variable environments, with relatively high salinities (UCANR 2017). Along the fluvial Petaluma River, hardy species such as Sacramento splittail would have tolerated the fluctuating temperatures and salinities associated with tidal influence and a variable flow regime. Steelhead/rainbow trout were

Figure 4.17. Sections of braided channels are visible in the 1942 aerial photos of Adobe Creek. The steep gradient of the Sonoma Mountains, high sediment loads, and substantial flow variability may have established this distinctive pattern. *(USDA 1942)*

Table 4.1. **Partial list of native plants associated with riparian corridors** that were historically present in the Petaluma River watershed, drawn from herbarium records. Where available, relevant information about the locality where the specimens were collected is included. Species were selected for inclusion in the table based on a combination of the locality information provided in the herbarium records and known associations with riparian types. All data were provided by the participants of the Consortium of California Herbaria.

Common Name	Scientific Name	Relevant Excerpt(s)	Year(s)	Notes
Annual hairgrass	Deschampsia danthonioides	Petaluma	1880	
Black hawthorn	Crataegus douglasii	Petaluma	1880	
California amaranth	Amaranthus californicus	San Antonio Creek	1939	
California blackberry	Rubus ursinus	Petaluma	1856?	Record has no date, but Emanuel Samuels spent a year collecting plants in California in 1856 (https://hdl.handle.net/2027/mdp.39015035491409)
California damasonium	Damasonium californicum	Petaluma	1880	
California grape	Vitis californica	Near Petaluma	1935	
California rose	Rosa californica	San Antonio Creek, "Petaluma Valley"	1893, 1910, 1939	
Carolina geranium	Geranium carolinianum	Petaluma	1924	
Cocklebur	Xanthium strumarium	Petaluma Valley	1893	
Common mouse tail	Myosurus minimus	Petaluma	1880	
Creeping leather root	Hoita orbicularis	Petaluma	1895	
Hairy gumweed	Grindelia hirsutula	Petaluma	1880, 1881, 1930	
Leafy pondweed	Potamogeton foliosus	San Antonio Creek	1939	
Meadow barley	Hordeum brachyantherum subsp. brachyantherum	Petaluma	1896	
Mule fat	Baccharis salicifolia	Petaluma	1921	
Narrowleaf willow	Salix exigua var. hindsiana	San Antonio Creek	1939	Listed in record as *Salix hindsiana*.
Oregon ash	Fraxinus latifolia	Burdell's School	1932	
Pink honeysuckle	Lonicera hispidula	Petaluma	1856?	Record has no date, but Emanuel Samuels spent a year collecting plants in California in 1856 (https://hdl.handle.net/2027/mdp.39015035491409)
Pink honeysuckle	Lonicera hispidula var. vacillans	Petaluma	1895	
Red willow	Salix laevigata	San Antonio Creek	1939	
Seep monkey flower	Mimulus guttatus	Petaluma	1880	Listed in record as Mimulus grandis.
Shade phacelia	Phacelia nemoralis subsp. nemoralis	Petaluma	1895	
Twinberry honeysuckle	Lonicera involucrata var. ledebourii	Petaluma	1895	
Water chickweed	Montia fontana	Petaluma	1933	
Western choke cherry	Prunus virginiana var. demissa	Hamilt(a)on Ranch. 1 mile from Old Adobe. East of Petaluma.		
Western dock	Rumex occidentalis	Petaluma	1880, 1890, 1921	

historically found in a number of tributaries within the watershed, including San Antonio Creek, Adobe Creek, Lynch Creek, Lichau Creek, and the Petaluma River mainstem (Leidy et al. 2005).

Within the Petaluma River mainstem, large in-channel pools likely provided summer refugia for native fish; three of the most well-documented pools are included in the historical synthesis mapping. Early land grant maps, for instance, show several pools near the head of the perennial river, the largest of which was referred to as the *Posas de Cantua* (Fig. 4.18). Oral histories by long-time Petaluma

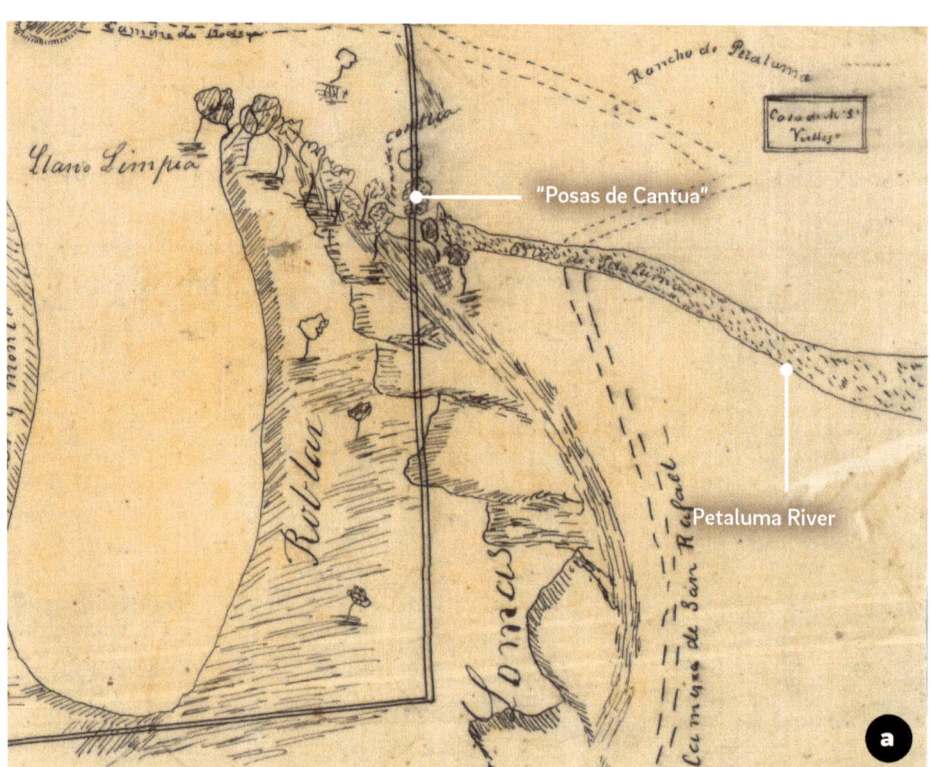

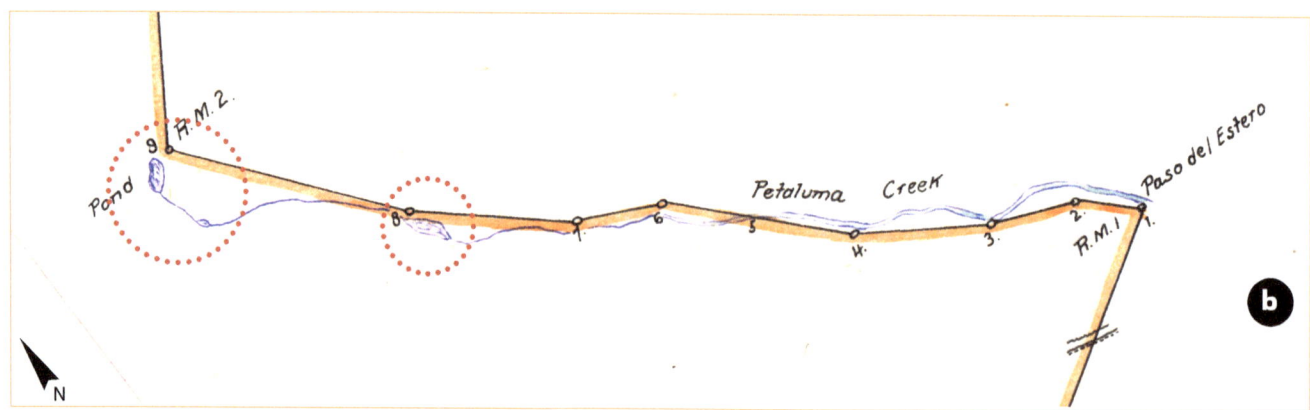

Figure 4.18. Early depictions of deep, in-channel pools in the mainstem Petaluma River. (a) An 1845 diseño labeled a chain of three pools "Posas de Cantua" at the headwater of the mainstem Petaluma River. (b) Three in-channel pools, the largest of which is labeled "pond," are represented on this 1857 plat of the Rancho Roblar de Miseria. *(a: USDC 1845, courtesy of The Bancroft Library; b: Thompson 1857c, courtesy of Bureau of Land Management)*

residents describe the presence of additional pools, or "swimming holes," further downstream. For example, resident John Stone remembered Big Green's and Little Green's holes near the output of Lynch Creek, and a big pool they called Ike's Hole close to the old Highway 101 bridge (Stone and Curme 1975; Fig. 4.19). As Stone recalled, the water ran through the deep pools, which were so full of fish—trout, splittails, and stickleback—that they used to catch "whole sacks of fish" (Stone and Curme 1975). Other accounts describe similar swimming holes under the Northwestern Pacific Railroad trestle, and a spot near the Cinnibar school in which the schoolchildren's grandparents swam in 1874 (Boivin 1998).

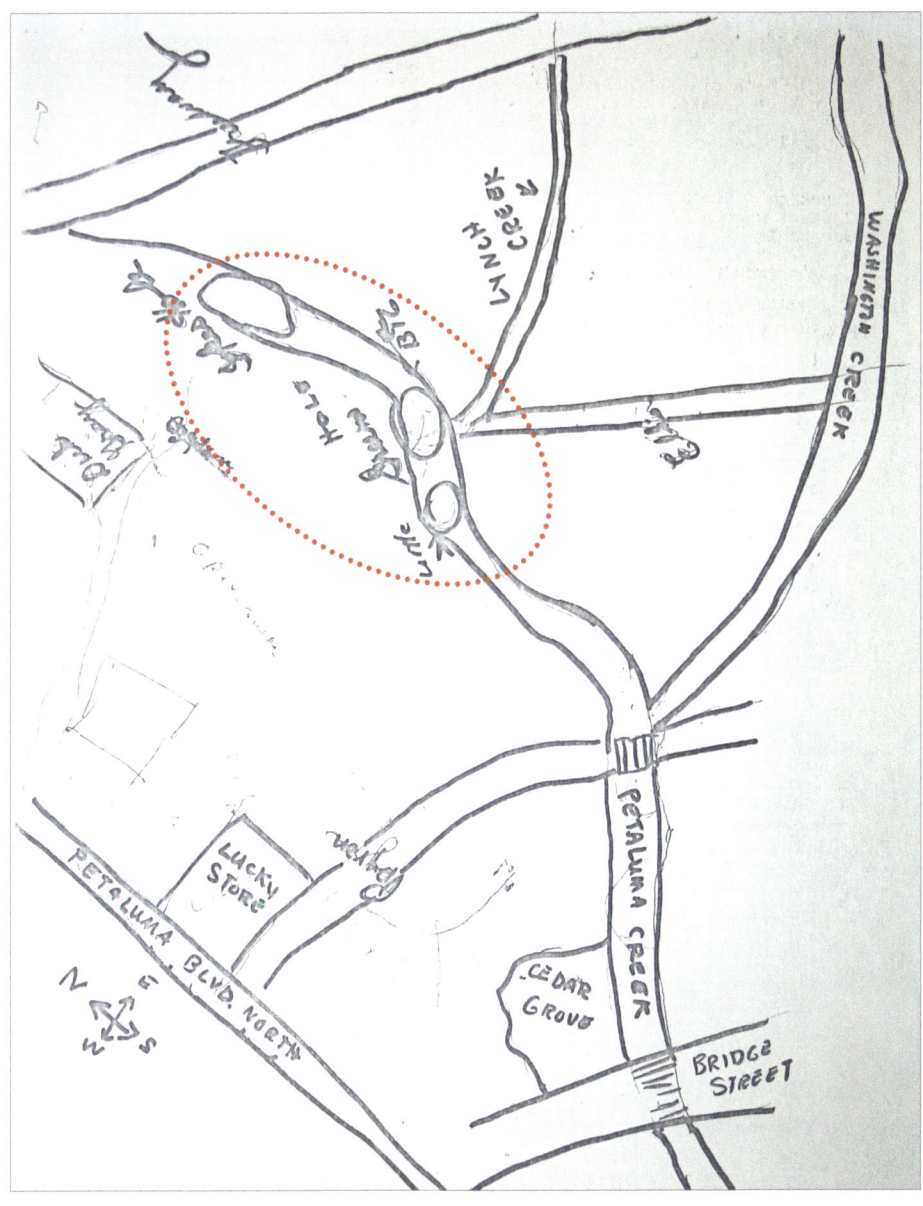

Figure 4.19. John R. Stone Sr. sketched the location of the swimming holes he remembers as a child. According to his account, these deep pools were full of fish. The water retained by in-channel pools may have provided important cold-water habitat for native fish, especially during the dry summer months. *(Stone and Curme 1975, courtesy of Sonoma State University Special Collections)*

5. NON-TIDAL WETLANDS

OVERVIEW

While the Petaluma River and its estuary were the focus of many early accounts and depictions of the landscape, non-tidal wetlands also occupied large swaths of the valley floor throughout the watershed in areas characterized by fine-grained soils, seasonal flooding, and high groundwater levels. Totaling roughly 4,610 ha (11,400 ac) these non-tidal wetlands provided habitat for many species of birds, mammals, and other wildlife. Seasonal wetlands, including wet meadow and vernal pool complex, maintained standing water or saturated soil for weeks to months during the wet season, and represented the majority of the non-tidal wetlands in the watershed (approximately 4,510 ha [11,150 ac]). Perennial wetlands, including valley freshwater marsh and willow groves, maintained saturated or flooded soils year-round, and were much more limited in extent (occupying just 110 ha [280 ac]).

This 1857 drawing looks southeast over the newly formed town of Petaluma. Ships are shown navigating along the Petaluma River, which is visible winding through the estuary south of town. Just beyond the river, the valley floor is depicted as a flat, treeless expanse extending to the base of the hills in the distance. *([VAULT] 917.94 K9, courtesy of California State Library)*

WET MEADOW

Wet meadow was by far the most extensive non-tidal wetland habitat within the watershed, occupying 4,120 ha (10,180 ac). Wet meadow is a seasonal wetland type composed of herbaceous plants such as grasses, sedges, rushes, and forbs. It typically occurs on poorly drained clay soils in areas with high groundwater or seasonal flooding, where water pools for extended periods during the wet season. Because of its poor drainage, landowners typically used wet meadow areas as pasture for livestock or for cultivation of lower-value, flood-tolerant crops such as hay or barley.

The largest expanse of wet meadow within the watershed was located on the flat valley floor on the northeastern side of the Petaluma River and tidal marsh (Fig. 5.1). It extended from the upper reaches of the estuary northeast to the base of the Sonoma Mountains, and from Ellis Creek northwest to the head of the Petaluma River. Traveling east from Petaluma in June of 1823, Spanish missionary Jose Altimira described this area as a "flat…covered with grass, but of little use for plants requiring irrigation in the summer; for in that season the springs are dried up, as is also the brook running on said flat or plain" (Altimira 1823). An early county history described the area as an "open waste of meadow land" (Munro-Fraser 1880).

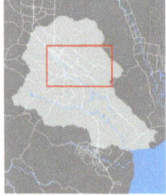

In the northern part of the watershed, wet meadow occupied large areas within the Lichau Creek, Willow Brook Creek, Liberty Creek, and Hutchinson/Wiggins Creek drainages. Patches of wet meadow also occurred through the San Antonio Creek drainage and along the margins of the tidal marsh.

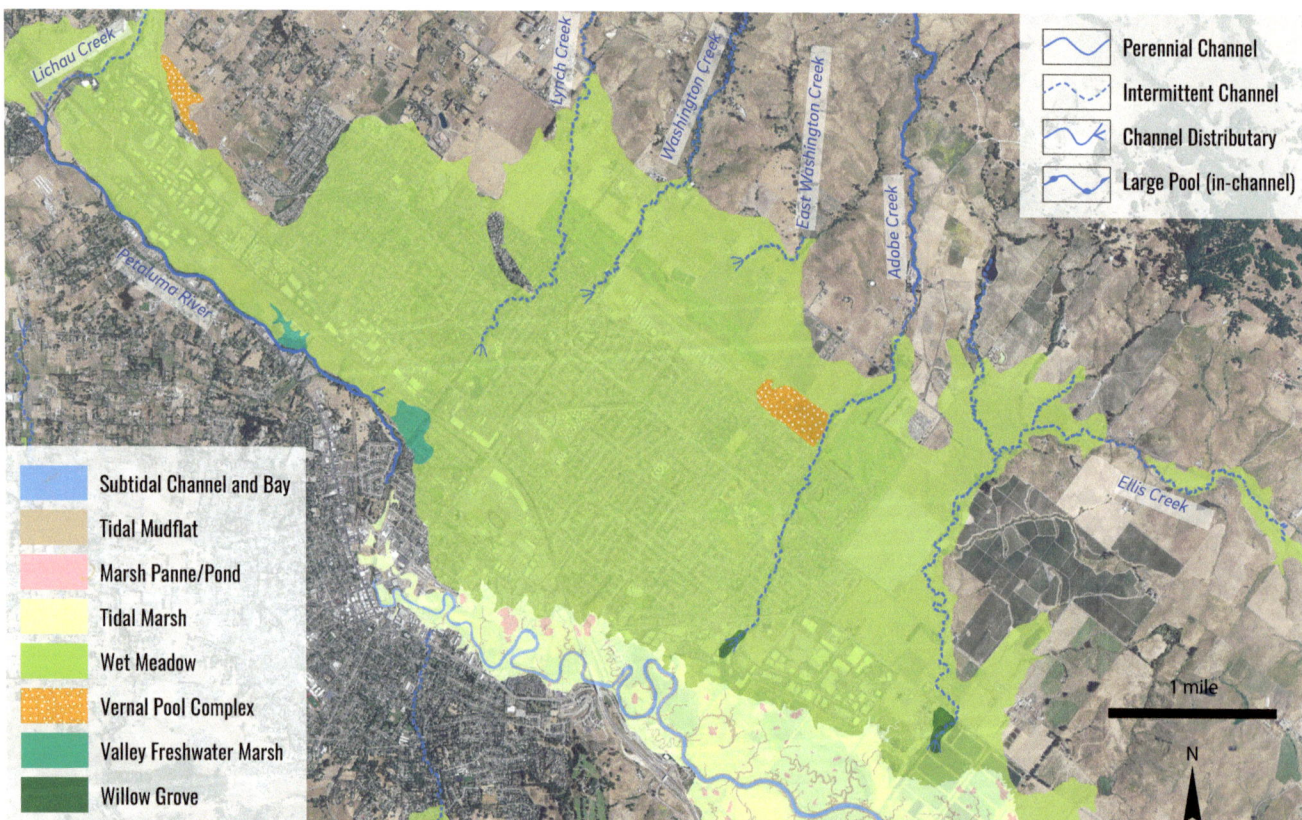

Figure 5.1. During the mid-19th century, thousands of acres of seasonally flooded wet meadow, with smaller patches of vernal pool complex and willow groves, occupied the valley floor east of the river. Tributaries draining off of the Sonoma Mountains dissipated as they flowed across the broad alluvial plain. *(NAIP 2016)*

The majority of the wet meadow in the watershed occurred on Dublin adobe soil, which is described in a 1914 soil survey as "more or less closely associated with sluggish drainage conditions… [where] water stands over large areas for days during the rainy season" (Holmes and Nelson 1914, 1917; Fig. 5.2). Patches of wet meadow also occurred in areas of Dublin loams and clay loams "along swales [where] drainage may be sluggish during the rainy season" (Holmes and Nelson 1914, 1917).

In addition to rainfall and seasonal flooding, springs, seeps, and high groundwater levels helped sustain seasonal wetlands in some locations. For example, approximately 200 ha (500 ac) of wet meadow and vernal pool complex occupied the Denman Flat basin where Liberty, Marin, Hutchinson, Wiggins, and Wilson creeks converged, near the head of the Petaluma River. While these wetlands were clearly supported in part by streamflow from these creeks, springs and seeps provided additional freshwater input: an 1845 diseño of Rancho Roblar de la Miseria shows several "ojitas de agua" (small springs) flowing from Meacham Hill into the wetland complex (Fig. 5.3). On the east side of the river between Petaluma and Lakeville, where the largest expanse of wet meadow occurred, the water table was reported to be approximately 12 feet below the surface in the late 19th century (Irelan 1893), though seasonal fluctuations likely brought the water higher during the rainy season. Groundwater levels had likely already declined in this area by the late 19th century as a result of groundwater pumping for agricultural use (Cardwell 1958). In addition, substantial stream channel incision had already occurred in some areas, such as along Ellis Creek, which may have further contributed to declines in groundwater levels on the plain (Irelan 1890).

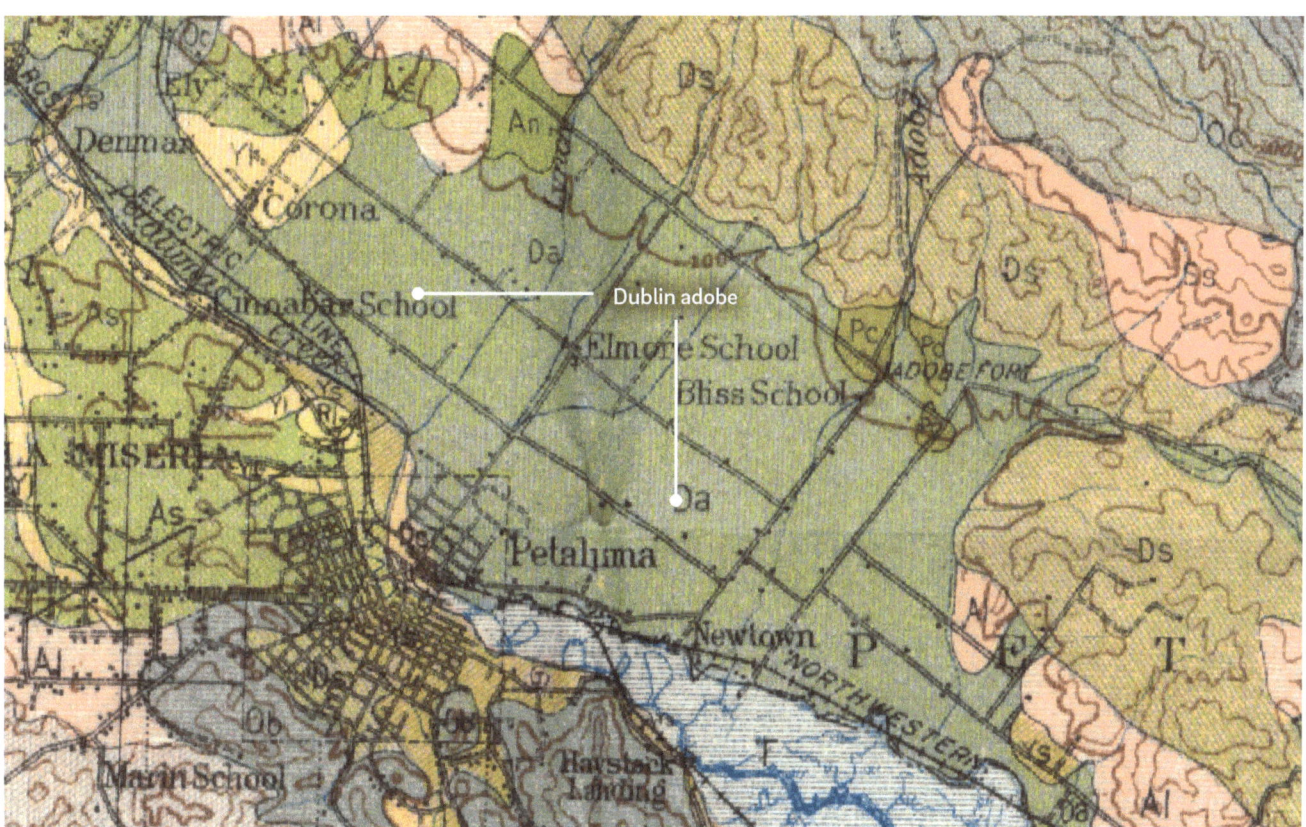

Figure 5.2. Wet meadow occurred on poorly drained clay soils such as Dublin adobe, which is shown in blue-green in this 1914 soil map. The accompanying report describes these soils as having "sluggish drainage conditions" (Holmes and Nelson 1917). The largest concentration of Dublin adobe occurred on the east side of the Petaluma River and estuary. *(Holmes and Nelson 1914, courtesy of USDA)*

Early botanical records from the vicinity of Petaluma include numerous accounts of species associated with wet meadows and other seasonal wetland types, such as sedges, meadow barley, Baltic rush, beardless wild rye, water chickweed, Pacific woodrush, Douglas' meadowfoam, peppergrass, and coastal button-celery (Table 5.1). Descriptions of the collection localities for these records, such as "marshy pasture," "wet valley fields," "meadow of grasses," "marshy pond," and "wet marsh and pools near Petaluma" are suggestive of wet meadow habitats. Seasonal wetlands within the watershed supported a number of plant species now designated as rare, threatened, endangered, or extirpated, including Petaluma popcornflower, pink star-tulip, Johnny nip, congested-headed hayfield, harlequin lotus, Pitkin marsh lily, Hickman's cinquefoil, Point Reyes checkerbloom, Suisun marsh aster, and two-fork clover (Fig. 5.4).

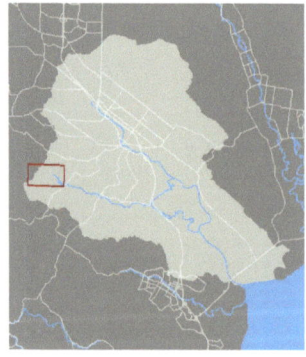

Figure 5.3. Seeps and springs ("ojitas de agua") on Meacham Hill (a) provided freshwater input that helped sustain wet meadows and vernal pool complexes within the Denman Flat basin downstream of Liberty, Marin, Hutchinson, and Wilson creeks (b). *(a: USDC 1845, courtesy of The Bancroft Library; b: NAIP 2016)*

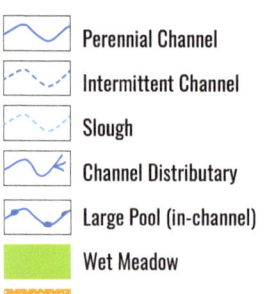

Figure 5.4. Type locality specimen of Petaluma popcornflower. This species is known from only one specimen, collected near Petaluma in 1880 by J.W. Congdon (Jepson 1928, NatureServe 2017). Presumed extinct today, this flower may have occurred in wet meadows, vernal pools, and possibly tidal marsh historically (Piper 1920, Kelley 2017, CNPS 2017). *(Barcode GH00093622, courtesy of The Harvard University Herbaria and the Botany Libraries)*

VERNAL POOL COMPLEX

Vernal pool complexes, totalling 390 ha (970 ac), were the second most extensive non-tidal wetland type within the watershed. These seasonal wetlands formed in topographic depressions underlain by an impermeable subsurface soil layer known as a hardpan, which inhibited drainage. Floodwaters and precipitation filled the depressions during the wet season, which then gradually dried out during the summer months.

Clusters of individual vernal pools, set within a grassland or wet meadow matrix, formed large complexes. On the northern side of the watershed, a large vernal pool complex occupied approximately 150 ha (370 ac) on the northwest side of Lichau Creek, just south of present-day Cotati. Vernal pool complexes occupied several hundred acres in the basin between the downstream ends of Liberty and Marin creeks, in the area around present-day Pepper Road and Rainsville Road. Several large vernal pool complexes were also found on the northern and southern ends of the Laguna wetland complex at the head of the San Antonio Creek drainage (Fig. 5.5; see pages 64–66). Additional smaller vernal pool complexes were found on the east side of the Petaluma River just south of Willow Brook Creek and within the large wet meadow expanse on the west side of Adobe Creek.

Figure 5.5. Vernal pools are visible near the head of San Antonio Creek in this ca. 1922 image, which is looking south towards the former Laguna de San Antonio (see pages 64-66). Spring Hill Road is in the foreground, and Chileno Valley Road (not visible) runs along the base of the hills in the background. Vernal pools and wet meadow surrounded the perennial wetlands at the core of the Laguna wetland complex. *(Dickerson 1922, Plate 24 [bottom], courtesy of Proceedings of California Academy of Sciences)*

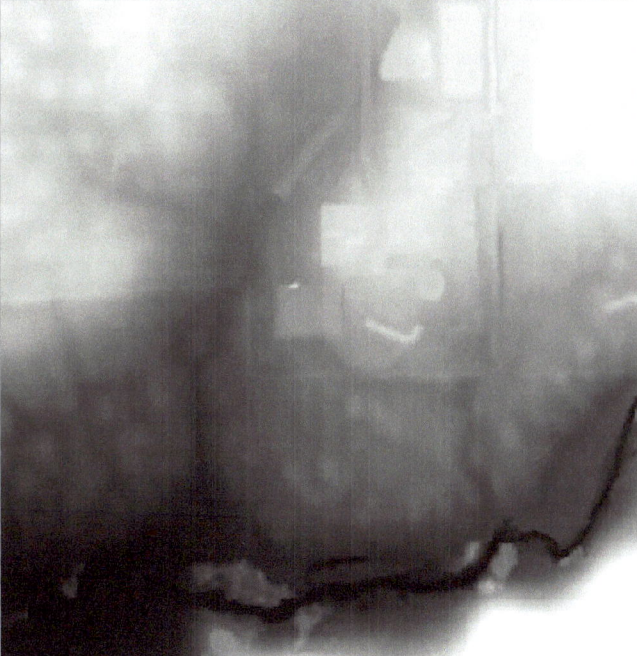

Figure 5.6. (above left) Vernal pool complexes appear as mottled areas in the 1942 aerial photographs. The individual pools are visible as darker patches, with intervening mounds visible as lighter areas. *(USDA 1942)*

Figure 5.7. (above right) Though many of the areas that historically supported vernal pools have been long-since graded over, remnants of the vernal pool "hog wallow" topography can still be discerned from modern LiDAR imagery. *(County of Marin 2015)*

N

500 feet

Many of the vernal pool complexes in the watershed occurred in areas of Madera loam, described in the early soil survey as "quite uneven, owing the presence of numerous small mounds and intervening depressions, also known as hog wallows" (Holmes and Nelson 1914, 1917). The surveyors note that "these depressions retain water during the rainy season." The hog wallow topography in these areas is visible in the 1942 aerial photographs, where the outlines of the mounds typically appear lighter than the depressions forming the individual vernal pools (Fig. 5.6). Remnants of this topography are also visible in modern LiDAR imagery (Fig. 5.7).

Vernal pools often support specialized plant and animal species adapted to their unique hydrological conditions. Botanical collections from the Petaluma area include numerous early records of species associated vernal pools or margins of vernal pools, such as common blennosperma, Johnny nip, California damasonium, yellow rayed goldfields, smooth tidy tips, stalked popcornflower, Oregon woolly marbles, and Lobb's aquatic buttercup (see Table 5.1).

Table 5.1. Partial list of native plants associated with perennial or seasonal non-tidal wetlands that were historically present in the Petaluma River watershed, drawn from herbarium records. Where available, relevant information about the locality where the specimens were collected is included. Species were selected for inclusion in the table based on a combination of the locality information provided in the

Habitat(s)	Common Name	Scientific Name	Relevant Excerpt(s)	Year(s)
W	Ajuga hedge nettle	*Stachys ajugoides* var. *ajugoides*	"Petaluma"	1880
M	Alkali marsh ragwort	*Senecio hydrophilus*	"Laguna in Chileno Valley," "marsh"	1946
W, V	Alkali milkvetch	*Astragalus tener* var. *tener*	"Petaluma"	1880
W, V	Annual checkerbloom	*Sidalcea calycosa*	"Petaluma"	1880
W, V	Annual hairgrass	*Deschampsia danthonioides*	"Petaluma"	1880
W, V	Annual semaphoregrass	*Pleuropogon californicus*	"Petaluma," "Between Petaluma and Chileno Valley"	1904, 1944
W, V	Annual semaphoregrass	*Pleuropogon californicus* var. *californicus*	"Petaluma," "Between Petaluma and Chileno Valley"	1880, 1944
W, M	Baltic rush	*Juncus balticus*	"Petaluma marshes"	1897
W	Bearded clover	*Trifolium barbigerum*	"Petaluma"	1880, 1904
W	Beardless wild rye	*Elymus triticoides*[1]	"Petaluma," "Roadside bank"	1892, 1897
W	Bitter cress	*Cardamine oligosperma*	"Petaluma"	1913
W	Bloomer's beaked buttercup	*Ranunculus orthorhynchus* var. *bloomeri*	"Petaluma"	1880
W, M	Bog yellow cress	*Rorippa palustris*	"Laguna (Chileno Valley)"	1947
W, V	Bracted popcornflower	*Plagiobothrys bracteatus*	"Petaluma"	ca. 1890?[2]
W	Bull clover	*Trifolium fucatum*		1928
M	Bur reed	*Sparganium eurycarpum* var. *greenei*[3]	"The Laguna, Chileno Valley"	1946
V	California damasonium	*Damasonium californicum*	"Petaluma," "Wet marsh and pools," "Pools; shallow shores"	1880, 1927, 1928
W	California hairgrass	*Deschampsia cespitosa* subsp. *holciformis*	"Petaluma"	1904
W	Coastal button-celery	*Eryngium armatum*	"Petaluma"	1888, 1903, 1930
W	Cocklebur	*Xanthium strumarium*	"Petaluma Valley"	1893
V	Common blennosperma	*Blennosperma nanum* var. *nanum*	"Petaluma"	1913
W	Common knotweed	*Persicaria lapathifolia*	"Laguna, Chileno Valley"	1947
W, V	Common mouse tail	*Myosurus minimus*	"Petaluma"	1880
W	Common tarweed	*Centromadia pungens* subsp. *pungens*	"West of Petaluma"	1931
W	Congested-headed hayfield tarplant[4]	*Hemizonia congesta* subsp. *congesta*	"4 mi nw Petaluma," "Six miles northwest of Petaluma," "On gentle slopes of low hills, in meadow of grasses," "From loose soil of bank"	1880, 1910, 1916, 1930, 1931
W	Cottontop	*Micropus californicus* var. *californicus*	"Petaluma"	1921
W	Creeping leather root	*Hoita orbicularis*	"Petaluma"	1895
W, M	Curvepod yellow cress	*Rorippa curvisiliqua*	"Laguna, Chileno Valley"	1947
W	Delphinium sp.	*Delphinium*	"Wet marsh and pools near Petaluma… in field"	1928
W	Douglas' meadowfoam	*Limnanthes douglasii* subsp. *douglasii*	"Petaluma"	1913, 1928
V	Douglas' silverpuffs	*Microseris douglasii* subsp. *tenella*	"Petaluma"	1921
V	Dwarf peppergrass	*Lepidium latipes*	"Petaluma"	1921
W, M	False waterpepper	*Persicaria hydropiperoides*	"Laguna in Chileno Valley"	1946
V	Forked pepperweed	*Lepidium oxycarpum*	"Petaluma"	1880, 1933
V	Fragrant fritillary	*Fritillaria liliacea*	"Petaluma"	1880
W	Harford sedge	*Carex harfordii*	"Petaluma"	1880
W	Harlequin lotus[5]	*Hosackia gracilis*	"Wet ground," "Petaluma"	1880, 1947
W, M	Hickman's cinquefoil[6]	*Potentilla hickmanii*	"Petaluma"	1880
M	Horned pondweed	*Zannichellia palustris*	"Marsh & pools near Petaluma," "Wet marsh and pools"	1928
W, V	Jepson's button celery	*Eryngium aristulatum* var. *aristulatum*	"3 miles northeast of Petaluma"	1933

herbarium records and known associations with non-tidal wetland types. The Habitat(s) column indicates the likely or possible wetland habitats for each species: W = Wet Meadow; M = Valley Freshwater Marsh; V = Vernal Pool Complex. All data were provided by the participants of the Consortium of California Herbaria.

Habitat(s)	Common Name	Scientific Name	Relevant Excerpt(s)	Year(s)
W, V	Johnny nip[7]	*Castilleja ambigua* subsp. *ambigua*	"Petaluma"	1880, 1889
W, V	Lemmon's canarygrass	*Phalaris lemmonii*	"Burdell Station"	1945
V	Lobb's aquatic buttercup[8]	*Ranunculus lobbii*	"Petaluma"	1928
W	Meadow barley	*Hordeum brachyantherum* subsp. *brachyantherum*	"Petaluma"	1896
W	Milk maids	*Cardamine californica*	"Two miles north of Petaluma," "In marshy pasture"	1940
W	Milk maids	*Cardamine californica* var. *integrifolia*	"Two miles north of Petaluma," "5 miles S of Petaluma," "Wet valley fields"	1929, 1940
W, M	Narrow manna grass	*Glyceria leptostachya*	"Petaluma," "Marshy pond 1 mile north"	1903, 1949
V	Oregon woolly marbles	*Psilocarphus oregonus*	"Petaluma"	ca. 1890?[9]
W, V	Pacific foxtail	*Alopecurus saccatus*	"Laguna, Chileno Valley"	1947
W	Pacific woodrush	*Luzula macrantha*	"Petaluma"	1870
W, V	Peppergrass	*Lepidium nitidum*	"Petaluma"	1913
W, V	Petaluma popcornflower	*Allocarya vestita*	"Petaluma"	1880[10]
W	Pink star-tulip[11]	*Calochortus uniflorus*	"Petaluma," "Wet marsh and pools"	1928
W, M	Pitkin marsh lily[12]	*Lilium pardalinum* subsp. *pitkinense*	"Petaluma"	1880
W, M	Point reyes checkerbloom[13]	*Sidalcea calycosa* subsp. *rhizomata*	"Petaluma"	1880
W	Rigid hedge nettle	*Stachys ajugoides* var. *rigida*	"Petaluma"	1895
W	Seep monkey flower	*Mimulus guttatus*[14]	"Petaluma"	1880
W	Slender sedge	*Carex gracilior*	"Petaluma"	1866
W, V	Slender woolly marbles	*Psilocarphus tenellus*	"Petaluma"	1921
W	Small-bracted sedge	*Carex subbracteata* var. *subbracteata*	"Petaluma"	1933
V	Smooth tidy tips	*Layia chrysanthemoides*	"Petaluma"	1921
W, M	Spike bent grass	*Agrostis exarata*	"Laguna, Chileno Valley"	1947
W, V	Stalked popcornflower	*Plagiobothrys stipitatus*	"Petaluma"	1933
W, M	Suisun marsh aster[15]	*Symphyotrichum lentum*	"Petaluma Marshes, near rail road track," "Laguna in Chileno Valley"	1897, 1921, 1946
W, M	Timothy canary grass	*Phalaris angusta*	"Laguna, Chileno Valley"	1947
M	Tule	*Schoenoplectus acutus* var. *occidentalis*	Laguna, Chileno Valley	1946, 1947
W	Two-fork clover[16]	*Trifolium amoenum*	"Near Petaluma," "Low rich fields"	1921
W, V	Water chickweed	*Montia fontana*	"Petaluma"	1933
W	Western dock	*Rumex occidentalis*	"Petaluma"	1880, 1890, 1921
W	Woolly goat chicory	*Macrorhynchus harfordii*	"Petaluma"	1870
V	Yellow rayed goldfields	*Lasthenia glabrata* subsp. *glabrata*	"Petaluma," "Burdell Station"	1933, 1945

[1] Listed in record as *Leymus triticoides*.

[2] No date given, but other collections by Edward Palmer in this location are ca. 1890.

[3] Listed in record as *Sparganium erectum* subsp. *stoloniferum*.

[4] CNPS Rare Plant Rank 1B.2

[5] CNPS Rare Plant Rank 4.2. Listed in 1880 record as *Lotus formosissimus*.

[6] CNPS Rare Plant Rank 1B.1

[7] CNPS Rare Plant Rank 4.2

[8] CNPS Rare Plant Rank 4.2

[9] No date given, but other collections by Edward Palmer in this location are ca. 1890.

[10] CNPS Rare Plant Rank 1A. Collection date appears to be mislabeled as 1850.

[11] CNPS Rare Plant Rank 4.2

[12] CNPS Rare Plant Rank 1B.1

[13] CNPS Rare Plant Rank 1B.2

[14] Listed in record as *Mimulus grandis*.

[15] CNPS Rare Plant Rank 1B.2. Listed in some records as *Aster lentus*.

[16] CNPS Rare Plant Rank 1B.1

VALLEY FRESHWATER MARSH

In addition to seasonal wetlands such as wet meadow and vernal pool complex, the watershed also supported approximately 100 ha (260 ac) of valley freshwater marsh. This perennial wetland type occurred in areas with fine-grained soils where standing water or saturated soil conditions persisted year-round.

Valley freshwater marsh formed the core of a large wetland complex at the headwaters of San Antonio Creek known historically as the "Laguna de San Antonio" (Fig. 5.8). A second laguna, today known as Laguna Lake, existed just over the drainage divide to the west in the Chileno/Walker Creek watershed. Mid-19th century diseños, the earliest known maps of these features, label the laguna at the head of San Antonio Creek as the "Laguna de San Antonio" (Fig. 5.9), and this laguna was the source of the Mexican land grant of the same name (Richardson 1853). However, the name was also occasionally applied to the laguna within the Chileno/Walker Creek drainage (Fig. 5.10).

Wetland habitats within the Laguna de San Antonio were distributed along topographic and hydrologic gradients, with drier, seasonal wetlands such as wet meadow and vernal pool complex located around the perimeter (see pages 56–63) and perennial valley freshwater marsh concentrated in lower elevation areas at the center. Perennial marsh areas likely consisted of a mix of emergent vegetation, dominated by tules, and open water pools and ponds that would have varied in size on a seasonal and interannual basis. For example, while some 19th century sources describe the Laguna as "tule marsh" or "fresh water tule land" (Lewis Publishing Company 1889, Unknown 1890), GLO surveyor Philip Thompson referred to it as a "pond of water" while conducting a survey of the Rancho Laguna de San Antonio in March of 1857 (Thompson 1857a); depictions of the Laguna in the early diseños also hint at this spatial and temporal variability (see Fig. 5.9). Several mid- to late 19th century county maps depict the Laguna using a series of concentric lines, further suggesting that flooding periodically transformed large areas into a seasonal lake (see Fig. 5.10). In addition to tules, early botanical records for freshwater marsh-associated species collected in this locality include spike

Figure 5.8. The Laguna de San Antonio, a wetland complex consisting of valley freshwater marsh, wet meadow, and vernal pool complex, occupied 260 ha (640 ac) at the head of San Antonio Creek. *(NAIP 2016)*

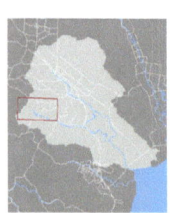

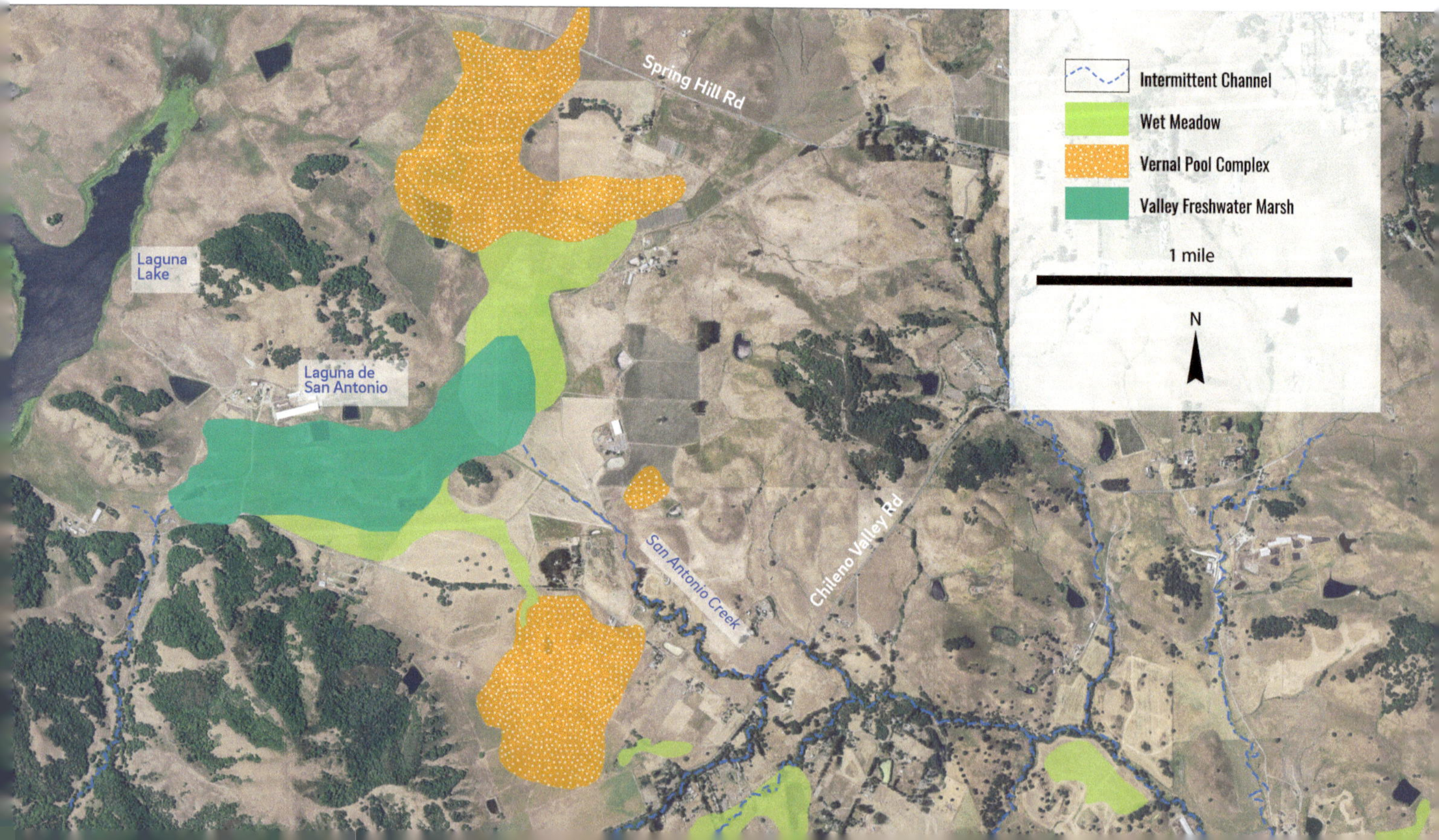

bent grass, false waterpepper, Timothy canary grass, curvepod yellow cress, bog yellow cress, alkali marsh ragwort, bur reed, and Suisun marsh aster (see Table 5.1).

Topographic controls contributed to the Laguna's setting at the head of San Antonio Creek. Bedrock outcroppings immediately east of the Laguna create a geologic constriction that would have impeded drainage into San Antonio Creek and contributed to the ponding of water within the upstream basin. Surface or subsurface inputs from the Laguna would have helped maintain baseflows in San Antonio Creek during the dry season (Collins et al. 2000). The earliest maps of the Laguna (Fig. 5.9) show it draining directly into San Antonio Creek, though the degree of connection with the creek would have varied seasonally and interannually, and during the driest months the connection would have been subsurface.

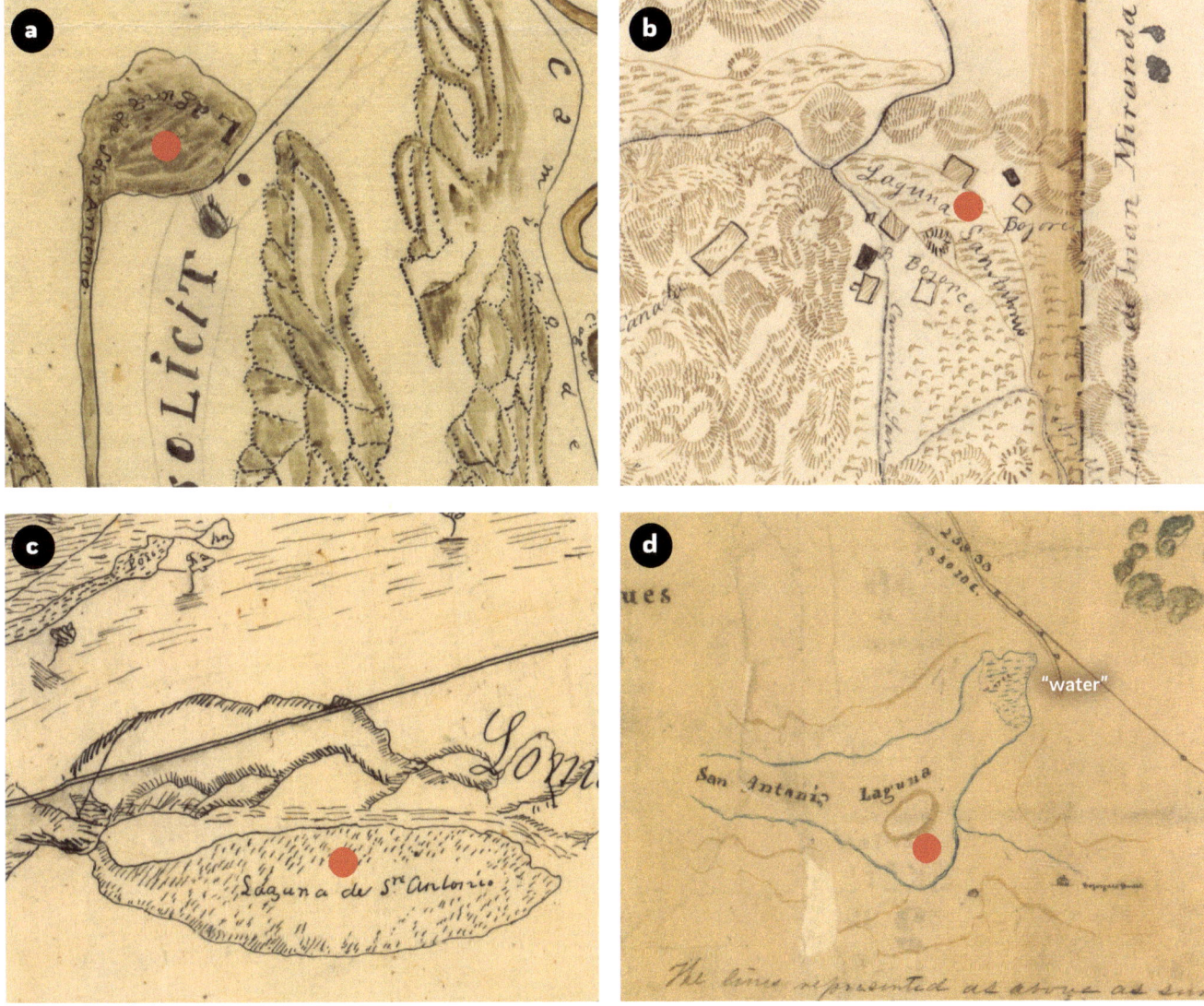

Figure 5.9. These diseños of Rancho Arroyo de San Antonio (a), Rancho Laguna de San Antonio (b), and Rancho Roblar de la Miseria (c,d), drawn ca. 1840, provide some of the earliest cartographic depictions of the Laguna de San Antonio. In two of the diseños (b,c), the Laguna is drawn with a conventional marsh or wetland symbol, suggesting that it was dominated by emergent vegetation such as tule. In one diseño (d), the northeastern portion of the Laguna is differentiated and labeled "water," while another (a) appears to depict the entire Laguna as an open water body. These differences hint at the spatial and temporal variability in the extent of flooding and open water habitat within the Laguna historically. Note that several diseños also show Laguna Lake, just over the drainage divide in the Chileno/Walker Creek watershed. *(a: USDC ca. 1856, courtesy of The Bancroft Library; b: USDC ca. 1844, courtesy of The Bancroft Library; c: USDC 1845, courtesy of The Bancroft Library; d: USDC 1852b, courtesy of The Bancroft Library)*

In addition to the perennial wetlands in the Laguna at the head of San Antonio Creek, additional pockets of valley freshwater marsh occurred along the northeastern side of the Petaluma River mainstem to the north of the present downtown area (Fig. 5.11). Freshwater marshes in the watershed likely supported a wide range of birds and other wildlife, including tricolored blackbird, a California Species of Special Concern (Mailliard 1879).

Figure 5.10. (above) Depictions of the Laguna de San Antonio in Sonoma and Marin county maps from 1866 (a) and 1873 (b), respectively. Note that these maps label Laguna Lake in the Walker/Chileno Creek drainage as Laguna de San Antonio, though the term was more commonly used to refer to the Laguna in the Petaluma River watershed. *(a: Bowers 1866, courtesy of David Rumsey Map Collection; b: Austin 1873, courtesy of David Rumsey Map Collection)*

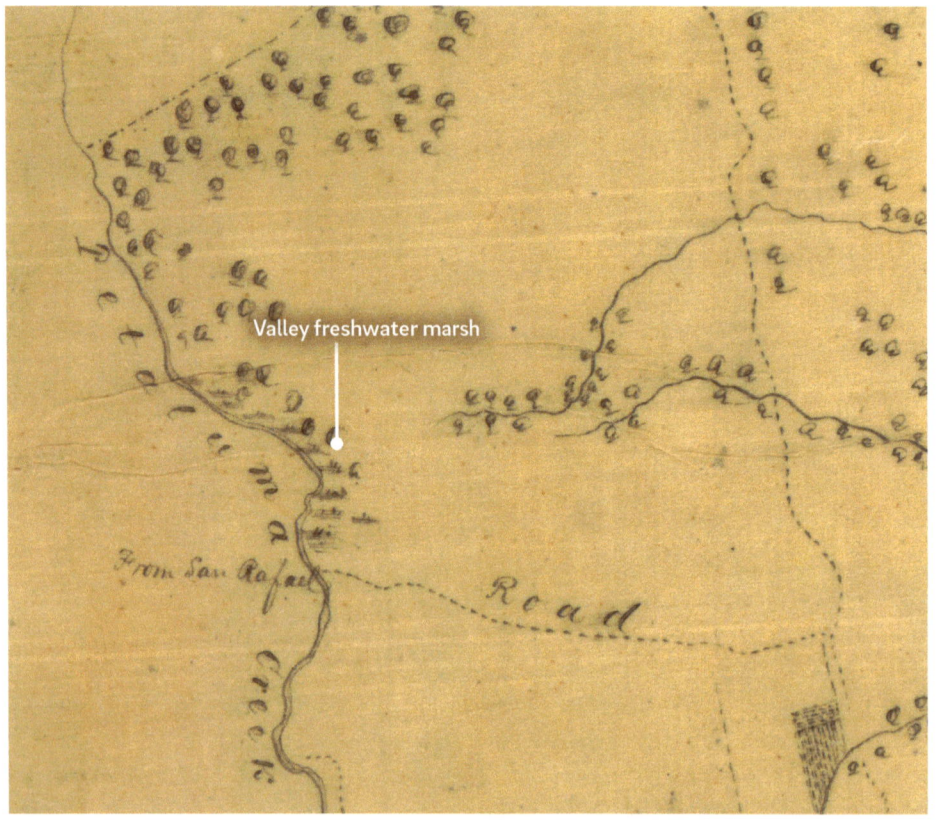

Figure 5.11. (left) This 1852 map of Rancho Petaluma shows several patches of perennial freshwater marsh alongside the mainstem of the Petaluma River a short ways upstream of the estuary. *(USDC 1852a, courtesy of The Bancroft Library)*

WILLOW GROVE

Several willow groves were found within the large wet meadow expanse on the eastern side of the valley, at the distal ends of Ellis and Adobe creeks, and occupied approximately 6 ha (15 ac). Unlike the riparian trees that bordered many of the creeks throughout the watershed (see pages 47–50), these clusters of willows, called "sausals" by early Spanish and Mexican settlers, were distinct features, and often occurred at the ends of intermittent streams or in areas with emergent groundwater, such as the base of alluvial fans (Collins and Grossinger 2004). Typically dominated by arroyo willow, sausals have been documented in similar landscape positions in many areas around the Bay, such as Livermore Valley (Stanford et al. 2013), Ygnacio Valley (SFEI-ASC 2016), and Santa Clara Valley (Grossinger et al. 2007, Beller et al. 2010). These wetlands provided valuable habitat for birds, amphibians, and other wildlife (Grossinger 2005, Grossinger et al. 2006).

The historical evidence for willow groves in the Petaluma River watershed is limited, especially in comparison to the well-documented groves elsewhere in the Bay Area. Their presence is inferred primarily from their landscape position in several mid-19th century maps, which depict the willow groves as distinct clusters of trees at the downstream ends of Ellis and Adobe creeks (Fig. 5.12). Zoologist Chester Lamb also reported finding an "extensive willow bottom 4 mi. S of town" while collecting specimens near Petaluma in September of 1927, though this may have been a reference to a particularly extensive patch of riparian forest along San Antonio Creek (Lamb 1927).

Figure 5.12. The 1852 map of Rancho Petaluma (a) shows several clusters of trees at the downstream ends of Adobe and Ellis creeks; based on landscape position, these are interpreted to represent depictions of small willow groves at the ends of the channels, just upstream of the estuary. Though USCS surveyors did not focus on landscape features beyond the extent of tidal influence, the 1860 T-sheet (b) also indicates the presence of a willow grove at the end of Ellis Creek. (a: USDC 1852a, courtesy of The Bancroft Library; b: Rodgers and Kerr 1860, courtesy of NOAA)

6. CHANGE OVER TIME

OVERVIEW

Over the past two centuries, agricultural and urban development (see pages 10-13) have resulted in major alterations to wetland and aquatic habitats and channels within the Petaluma River watershed. These changes include wetland loss, conversion from one wetland type to another, introduction of novel wetland types, degradation of habitat quality, and channel modifications (Fig. 6.1 and Fig. 6.2).

Overall, approximately 68% of wetland and aquatic habitat area has been lost (Fig. 6.3). Among tidal wetland types, habitat extent decreased by 58%.[1] Among non-tidal wetland types, habitat extent decreased by 84%.[2] Major changes have also occurred within the channel network. Many streams have been lengthened and channelized in order to increase

---

[1] This calculation includes historical tidal wetlands types (Tidal Marsh, Tidal Mudflat, Subtidal Channel, Marsh Panne/Pond) as well as tidal unnatural lagoon features.

[2] This calculation includes historical non-tidal wetland types (Valley Freshwater Marsh, Wet Meadow, and Vernal Pool Complex; Willow Grove and Large Pool were excluded) as well as novel non-tidal wetland types that were not present historically (Depressional Vegetated, Depressional Open Water, and non-tidal Unnatural Lagoon features).

Photo by Greenbelt Alliance, April 2012, licensed under Creative Commons

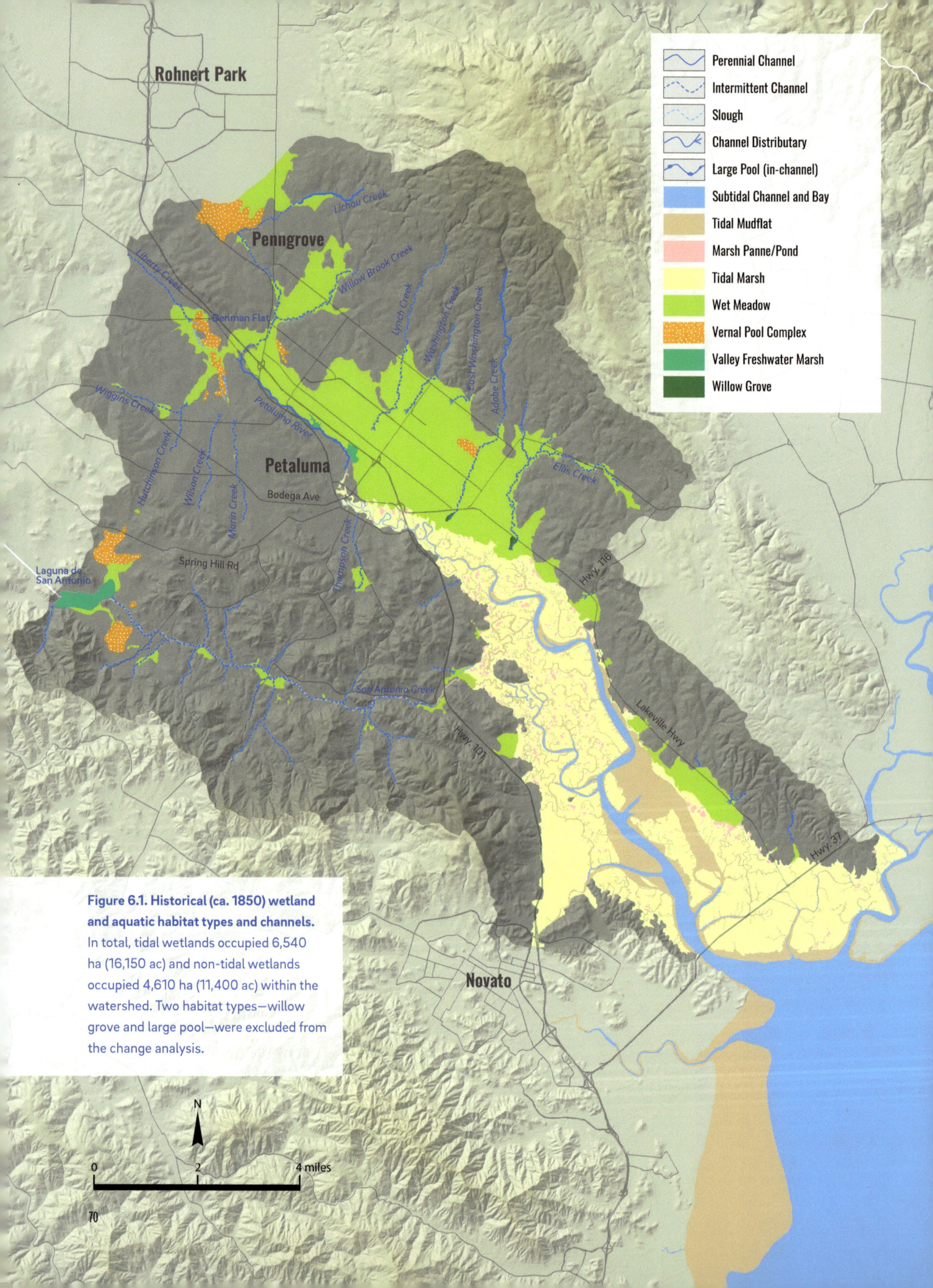

Figure 6.1. Historical (ca. 1850) wetland and aquatic habitat types and channels. In total, tidal wetlands occupied 6,540 ha (16,150 ac) and non-tidal wetlands occupied 4,610 ha (11,400 ac) within the watershed. Two habitat types—willow grove and large pool—were excluded from the change analysis.

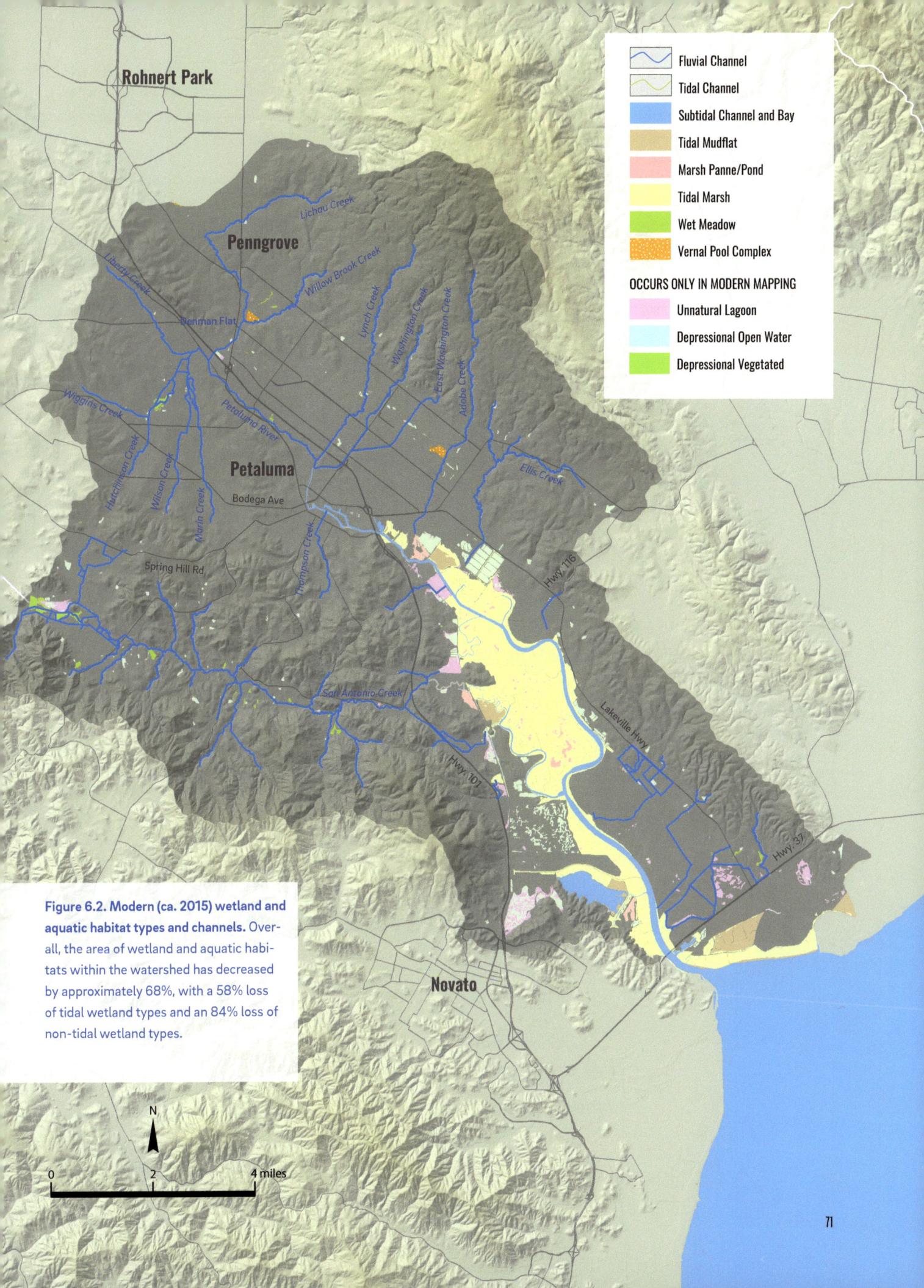

Figure 6.2. Modern (ca. 2015) wetland and aquatic habitat types and channels. Overall, the area of wetland and aquatic habitats within the watershed has decreased by approximately 68%, with a 58% loss of tidal wetland types and an 84% loss of non-tidal wetland types.

drainage efficiency, which has resulted in an approximately 50% increase in channel length among the largest tributaries. Channel straightening and other changes in channel alignment have also taken place throughout the watershed, most notably at the mouth of San Antonio Creek.

This chapter discusses the major changes in wetland extent and distribution and channel configuration that have occurred over the past century and a half. See page 20 for a description of the methodology and data sources used in the landscape change analysis.

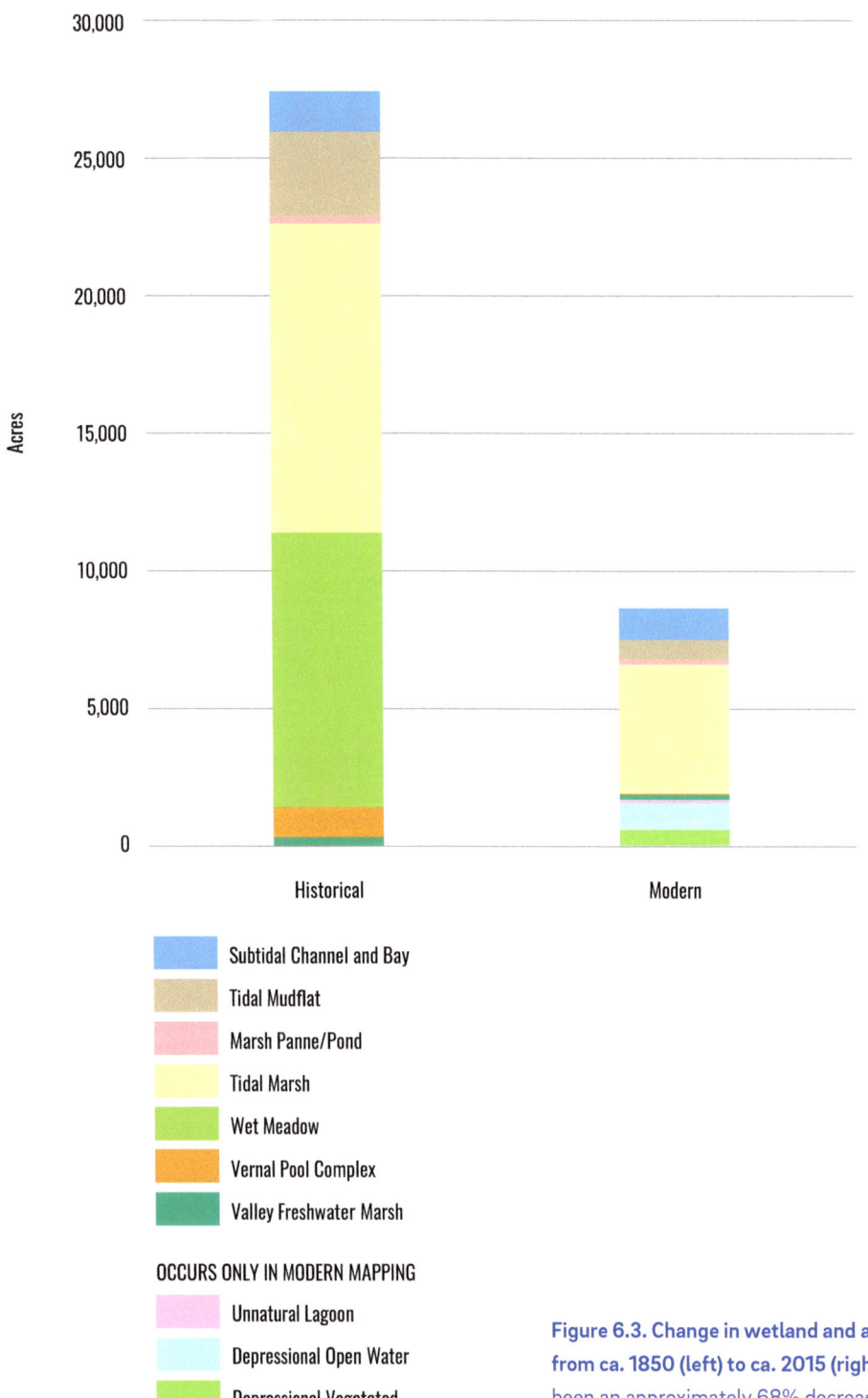

Figure 6.3. Change in wetland and aquatic habitat area from ca. 1850 (left) to ca. 2015 (right). Overall there has been an approximately 68% decrease in habitat extent.

TIDAL MARSH DIKED, DRAINED, AND FILLED

Tidal wetland area within the watershed has decreased by 58%, from approximately 6,540 ha (16,150 ac) in 1860 to 2,760 ha (6,810 ac) today. The loss of tidal wetlands was well underway by the mid- to late 19th century, as large areas of marsh were diked and drained and converted to agricultural land uses, particularly on the southeastern side of the river (Fig. 6.4, Fig. 6.5; Heuer 1917). Writing in 1888, for instance, USCGS surveyor James Lawson noted the extent to which the marshes around the mouth of the Petaluma River were already under cultivation:

> From Tolay Cr. Westward to Petaluma Cr. And along the Eastern shore of the latter, a large amount of marsh has been reclaimed by dykes and ditches, and is successfully cultivated...On Western side of Petaluma Creek, 1 ½ miles inside of Entrance, a considerable portion of the marsh between the shoreline and fast land has been reclaimed, and part of it cultivated. (Lawson 1886-7)

Former baylands were used to grow a variety of moisture-tolerant crops, such as oats, hay, barley, and sugar beets (*Petaluma Courier* 1887a, *Petaluma Courier* 1887b, *San Francisco Call* 1900, *San Francisco Call* 1913), and were also used for dairying (Heuer 1917).

Other land uses also contributed to the loss of tidal wetlands during the 19th and early 20th centuries. Several railroad lines, for instance, were constructed through the marsh and across the Petaluma River during the late 1800s: the San Francisco &

Figure 6.4. By the late 19th century, large areas of former tidal wetlands had been "reclaimed" and brought into agricultural use, as can be seen in this 1887 USCGS T-sheet. Levees are shown surrounding the newly cultivated areas on both the east and west sides of Petaluma Creek. Traces of former tidal channels are still visible in many areas within the diked baylands. *(Lawson and Welker 1887, courtesy of NOAA)*

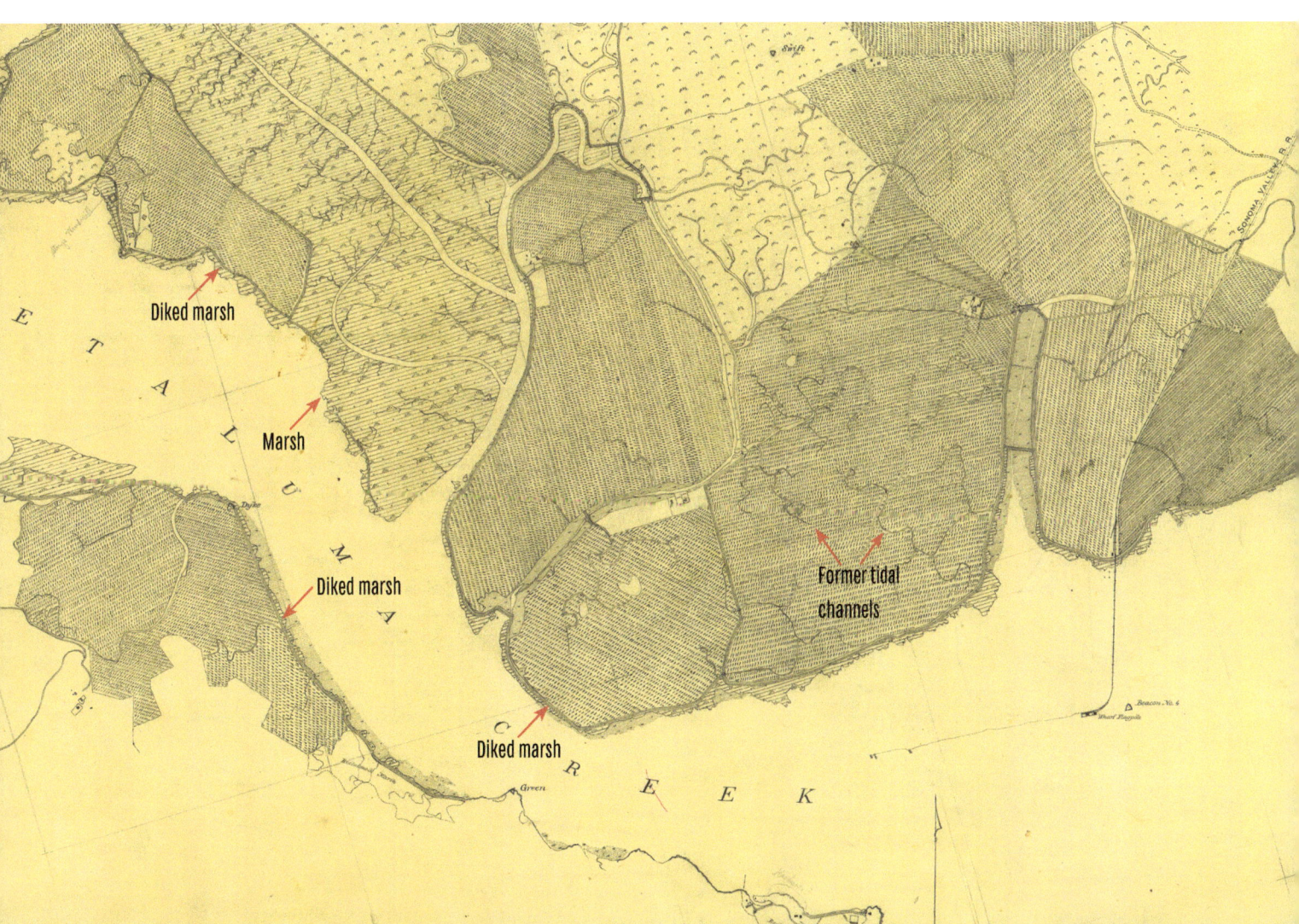

Figure 6.5. This 1871 map shows a "flood gate" and "dam" regulating tidal flows within Duncan Slough, just east of the Petaluma River mouth. The SF&NPRR line (originally part of the Marin and Napa Rail Road Company) was not constructed until 1887-8, and was apparently added on to the map at a later date. *(Thompson 1871, courtesy of Curtis & Associates, Inc.)*

North Pacific Railway Company line ran along the western side of the marsh and crossed the river just southeast of downtown Petaluma (the original bridge was completed in 1880), while the Marin and Napa Rail Road Company line (constructed in 1887-8) crossed through the marsh and over the river approximately 0.75 km (0.5 mi) upstream of San Pablo Bay (Stindt and Dunscomb 1964, Heig 1982). Fill and pilings used to construct the railroad lines resulted in wetland loss, and also restricted the flow of water within the marsh. A number of other major infrastructure projects, including portions of Highway 37, the Marin County Airport, Redwood Landfill, and the Ellis Creek Water Recycling Facility, have been constructed on former baylands.

False Bay, which was a large tidal mudflat at the time of the earliest USCS surveys, has experienced several distinct changes over the past 150 years. By the early 20th century, sediment accretion had caused much of the mudflat at False Bay to be converted to vegetated marsh. Though still referred to as a "high water lagoon" as late as 1917 (Heuer 1917), by 1922 UCSGS surveyor O.W. Swainson reported that "False Bay has filled up to a great extent" (Swainson 1922). The area remained as an undiked marsh for a period during the 1920s and 30s, though by 1942 much of the area was surrounded by levees and was being brought into cultivation (Fig. 6.6).

Despite the substantial loss of tidal wetland area, the Petaluma Marsh still comprises the largest remaining ancient tidal marsh in San Pablo Bay, providing habitat for rare, threatened, and endangered species such as the California clapper rail, black rail, and salt marsh harvest mouse (SRCD 2015). In addition, a number of conservation and restoration efforts have begun to bring back some of the "reclaimed" tidal wetlands, or enhance degraded wetlands, in areas such as the Sonoma Baylands, Sears Point, Carl's Marsh, Rush Creek, Bahia Marsh, and Shollenberger Park (SFEP 2014, SFEI 2017).

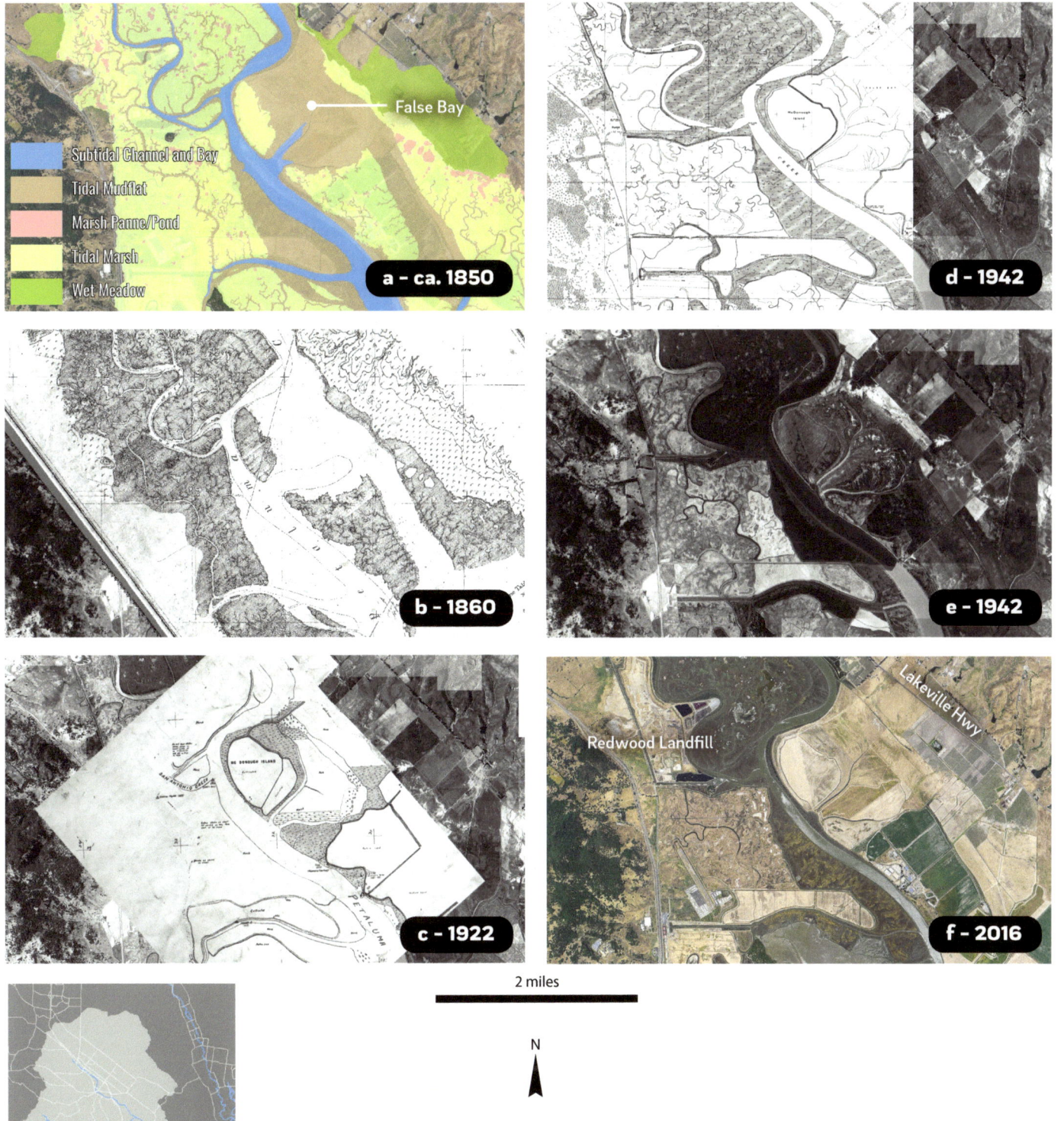

Figure 6.6. This sequence of images shows the gradual filling of False Bay over the late 19th and early 20th centuries. Historically a shallow tidal mudflat (a,b), by 1922 sediment accretion had caused much of the area to be converted to vegetated marsh (c). By 1942, much the area had been leveed (d,e), but was not yet heavily cultivated as it is today (f). *(a: NAIP 2016; b: Kerr 1860, courtesy of NOAA; c: Swainson et al. 1922; d: USCGS 1941-2, courtesy of NOAA; e: USDA 1942; f: NAIP 2016)*

PETALUMA RIVER DREDGED AND STRAIGHTENED

Since as early as 1860, efforts have been made to make the Petaluma River more conducive to maritime navigation through channel dredging and straightening within the tidal reaches. The first documented channel modifications, during the 1860s, resulted in the excavation of several cut-offs in the reach around Haystack Landing (*Sonoma County Journal* 1862, Roop and Flynn 2007).

In 1880, the Army Corps initiated its first large-scale dredging operation in the river, which created three cut-offs, deepened the river to 1 m (3 ft) at low tide, and widened it to 15 m (50 ft) (Schulz 1927; Fig. 6.7). Additional dredging, straightening, and maintenance occurred during the late 19th and early 20th centuries, including the creation of McNear Canal in 1893 and the dredging of the turning basin from 1918–21 (Heuer 1917, Schulz 1927, Roop and Flynn 2007). Another major Army Corps dredging project was conducted in 1931, which increased the channel's width to 30 m (100 ft) and its depth to 2 m (8 ft) from the river mouth up to Western Avenue (United States Army War Department 1933). Dredged sediment was often used to fill parts of the surrounding tidal marsh or former tidal channels (Mendell 1883, Stone and Curme 1975; Fig. 6.8). In an interview conducted in 1975, for instance, John Stone, a long time resident of Petaluma, described how dredged sediment was used to fill old channels in the 1920s and 30s:

> [Curme]: They filled in this old creek bed? [Stone]: Afterwards. Well, they didn't do it until years later. They pumped in a little at a time. They put in bridges across out here so they could get out there...[Curme]: Then all this—the original creek bed—is now man-made, or artificial, land? [Stone]: Yes, that's all artificial, that whole thing back there. (Stone and Curme 1975)

Figure 6.7. During the late 19th and early 20th century, numerous "cut-offs" were excavated at bends in the Petaluma River in order to facilitate maritime navigation. This 1931 Army Corps map shows two of the cut-offs, as well as the location of the "old Petaluma Creek bed." *(Gonzalez 1931, courtesy of Curtis & Associates, Inc.)*

Figure 6.8. The caption of this 1937 photograph, looking north from the top of the grain elevator on East Washington Street, says, "Dredge pipe line delivering a fine sand from the bottom of the Petaluma River to low land in the north." The dredge pipe is visible just beyond the railroad bridge over the Petaluma River, and appears to be depositing the sediment in a low area that was a former meander of the river. *(Annex photo 33380, courtesy of Sonoma Heritage Collection)*

Sediment accumulation within the river necessitated periodic dredging on an ongoing basis (Fig. 6.9). Between 1908 and 1927, for instance, an estimated 28,290 m^3 (37,000 yd^3) of sediment were removed from the channel annually above Haystack Landing in order to maintain existing channel dimensions (Jackson 1927). Increased sediment accumulation within the Petaluma River channel was the result of multiple factors, including erosion stemming from grazing, farming, and other land uses within the watershed; loss of tidal prism due to diking and filling of the surrounding marsh; and the influx of sediment from hydraulic mining following the Gold Rush (Jackson 1927, USACE 1974, Atwater et al. 1979). Sediment accumulation resulted not only in decreased channel depth, but also progradation of the marsh at the mouth of the river and along the margins of the channel. Marsh progradation rates at the mouth of the river averaged 1–2 mm/yr (0.04–0.08 in/yr) between ca. 1855 and 2010 (Beagle et al. 2015), and in the lower reaches of the river the width of the subtidal channel and tidal mudflat has decreased by up to 600 m (2,000 ft) over the past 150 years (Fig. 6.10).

To maintain and enhance the navigability of the channel, dredging and straightening efforts continued through the 20th century (SCWA 1986). The cumulative modifications to the river have significantly altered channel planform, decreasing the sinuosity of the tidal portions of the Petaluma River from approximately 4 cm (1.5 in) the mid-19th century to 5 cm (1.2 in) today (Fig. 6.11).

In the year 1850, when the township was first commencing to be settled, the depth of the creek was considerably greater than it is to-day; debris had not yet been cast into its clear waters nor had mud formed in such vast quantities on its banks.

—Munro-Fraser 1880

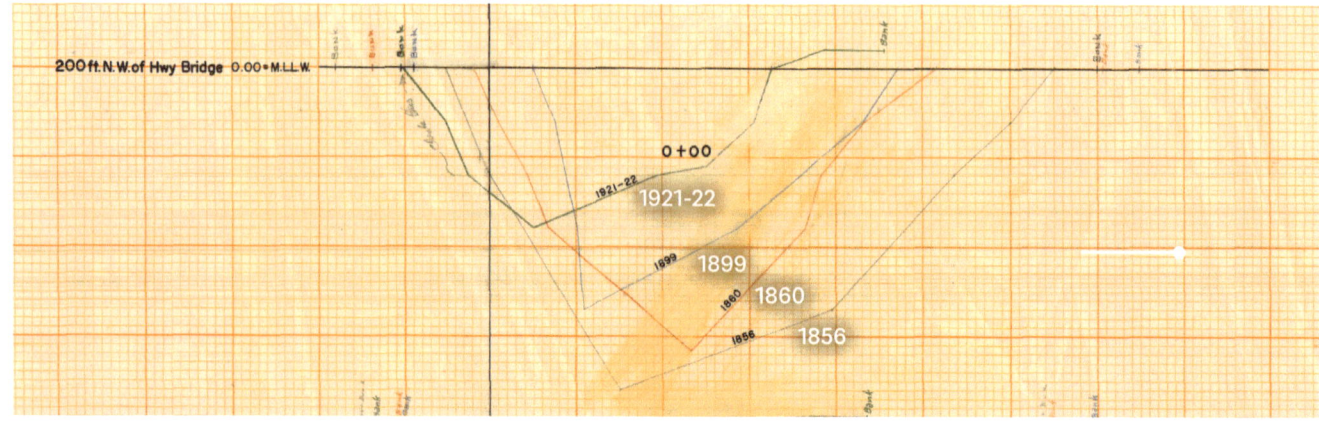
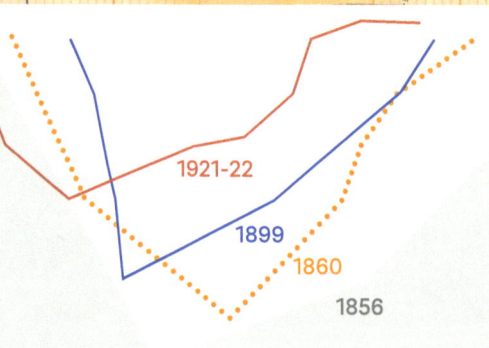

Figure 6.9. (above) The effects of sediment accumulation on channel elevations in Petaluma Creek are apparent in this USCGS chart showing cross-sections taken near Black Point in 1856, 1860, 1899, and 1921-2. The maximum channel depth, which was over 11 m (35 ft) in 1856, had decreased to less than 6 m (20 ft) by 1921-2. *(USCGS 1952, courtesy of State Lands Commission)*

Figure 6.10. (left) Comparison of the 1854 T-sheet (a), the 1942 aerials (b) and 2016 NAIP imagery (c) illustrates the progradation of tidal marsh on the margins of the Petaluma River channel. Sediment accretion resulting in conversion to vegetated marsh has decreased the width of former mudflats and tidal channels near the mouth of the river by up to 600 m (2,000 ft). *(a: Rodgers 1854, courtesy of NOAA; b: USDA 1942; c: NAIP 2016)*

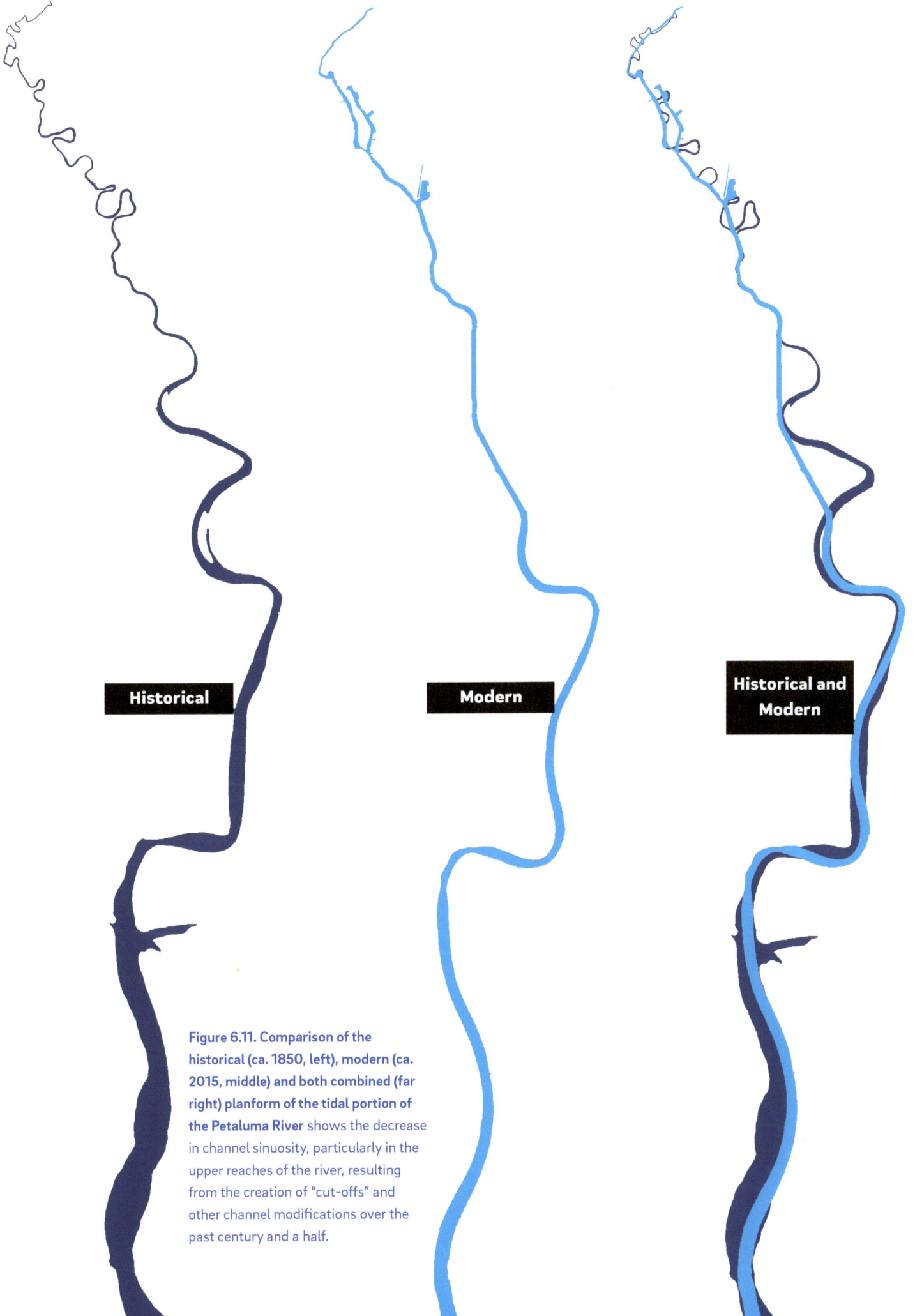

Figure 6.11. Comparison of the historical (ca. 1850, left), modern (ca. 2015, middle) and both combined (far right) planform of the tidal portion of the Petaluma River shows the decrease in channel sinuosity, particularly in the upper reaches of the river, resulting from the creation of "cut-offs" and other channel modifications over the past century and a half.

LOSS OF SEASONAL WETLANDS

In addition to the loss of tidal wetlands and modifications to the Petaluma River, land use changes have resulted in major transformations to non-tidal wetlands throughout the watershed, including seasonal wetlands such as wet meadow and vernal pool complex. Across the watershed, wet meadow area has declined by 98% (from 4,120 ha [10,180 ac] to 80 ha [190 ac]), while vernal pool complex has declined by 95% (from 390 ha [970 ac] to 20 ha [50 ac]).

Virtually all of the large, contiguous expanse of wet meadow that historically existed on the northeastern side of the river has been eliminated. The initial loss of this habitat was largely due to conversion to agricultural land uses (Fig. 6.12a). Today, however, much of this area is dominated by urban development (Fig. 6.12b). Small remnants of wet meadow exist in a number of locations throughout the watershed, most notably at the head of San Antonio Creek, along the San Antonio Creek mainstem and tributaries, and along Wiggins Creek.

In addition to direct impacts from urban and agricultural development, declines in groundwater levels may have also contributed to the loss of wet meadow and other non-tidal wetlands throughout the watershed. Historically, wet season groundwater levels would have been close to the surface in many of the areas supporting wet meadow, such as the valley floor on the northeast side of the river. By the mid-20th century, however, groundwater levels in the valley had fallen to 3–8 m (10–25 ft) below the surface; during the dry season the water table declined still further, to depths of 12 m (40 ft) or more (Cardwell 1958). Groundwater levels recovered somewhat during the 1960s and 70s, as the municipal water supply shifted to greater reliance on imported water and rates of groundwater pumping declined (DWR 1982), though recent data from water supply wells indicates that groundwater levels in many areas that historically supported non-tidal wetlands are still at least 3 m (10 ft) deep (and in some cases much deeper; https://

Figure 6.12. Historical wet meadow habitat on the valley floor east of the Petaluma River (indicated in green) was initially converted to agricultural land uses (a), though today this area is dominated by urban development (b). *(a: USDA 1942; b: NAIP 2016)*

Figure 6.13. Though 95% of historical vernal pool extent has been lost, small remnants of vernal pool habitat exist in several locations throughout the watershed, including patches adjacent to Adobe Creek (a) and Willow Brook Creek (b). *(NAIP 2016)*

geotracker.waterboards.ca.gov/; http://www.water.ca.gov/waterdatalibrary/). Contemporary data on groundwater levels in the watershed is limited, though a USGS-led groundwater study is currently in progress (https://ca.water.usgs.gov/projects/2012-02.html).

Most of the vernal pools that existed within the watershed historically have been eliminated by urban and agricultural development, though small remnants of vernal pool complex still exist on the valley floor just west of Adobe Creek and near Willow Brook Creek (Fig. 6.13). In a number of areas where land use and hydrologic changes have destroyed or degraded vernal pool habitat and ecological function, the topographic features that supported vernal pools historically—mounds and depressions—are still present in some form and can be discerned in LiDAR imagery (see Fig. 5.7 on page 61).

DRAINING OF LAGUNA DE SAN ANTONIO

By 1880, the Laguna wetland complex at the head of San Antonio Creek was ditched and drained, and former wetland areas had been brought into agricultural use (Alley, Bowen & Co. 1880; Fig. 6.14). Wheat, barley, and potatoes were among the first crops to be cultivated on the newly drained lands (*Marin Journal* 1878; Alley, Bowen & Co. 1880; *Petaluma Weekly Argus* 1886), which were also subsequently used to cultivate hay and corn and as pasture for dairy cows (Dolcini Family 1984, 2006, 2007). Despite being ditched and drained, the area continued to provide some function as aquatic habitat: a *Marin Journal* article from February of 1901 reported that "the upper and lower laguna are well filled with water since the last rains and duck hunting is quite the rage" (*Marin Journal* 1901). Today, depressional and seasonal wetlands still occupy approximately 55 ha (140 ac) at the head of San Antonio Creek, within the footprint of the historical Laguna, though these wetlands are highly modified, and do not provide habitat quality comparable to the large wetland complex that existed in the past (Fig. 6.15).

The loss of the Laguna wetlands has likely had a significant impact on the hydrology of San Antonio Creek. Collins et al. (2000) hypothesize that the historical Laguna acted as a natural reservoir, storing flood waters in the winter and releasing them during the dry season. The loss of the Laguna would have impacted both of these functions, resulting in both greater peak flows and decreased base flows in San Antonio Creek. Along with other land use changes in the San Antonio Creek watershed that have contributed to soil erosion, the increase in peak flows in San Antonio Creek is likely a major factor contributing to bed incision and channel erosion within the upper and middle reaches of San Antonio Creek (Fig. 6.16, Fig. 6.17).

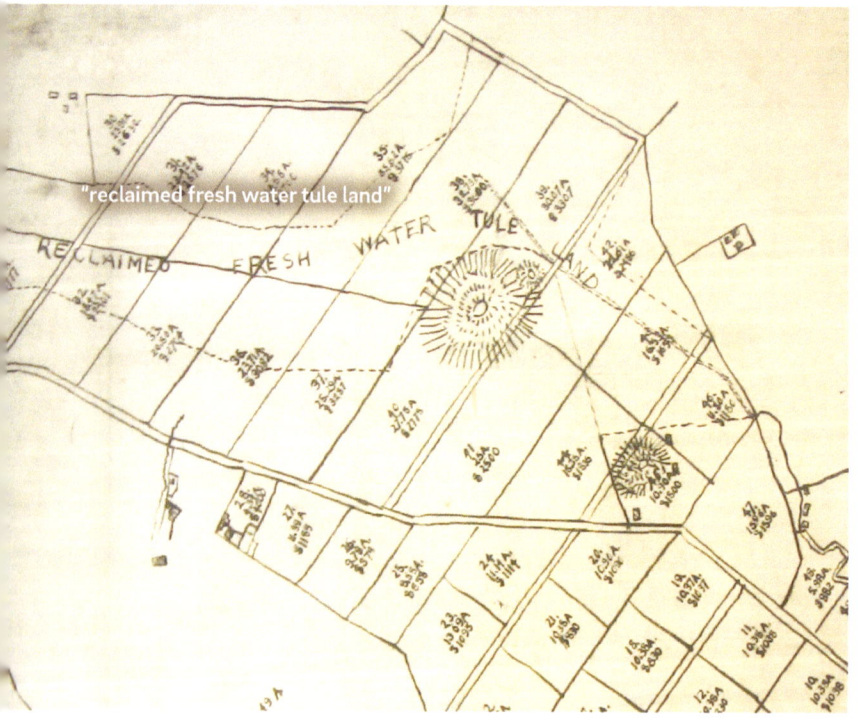

We found L.W. Walker deepening and widening his ditch which drains and reclaims the San Antonio Laguna. Further on we found the laguna, the waters of which used to flow down through the Chileno Valley, now covered with a heavy crop of barley just getting ripe. These barley fields used to be covered with water mid-summer and was the favorite resort of sportsmen in quest of ducks and geese.
—Petaluma Weekly Argus 1886

Figure 6.14. **The Laguna de San Antonio was ditched and drained in the late 19th century** and used for agricultural production. This 1890 map of the "Walker Villa Tract" labels the Laguna as "reclaimed fresh water tule land." *(Unknown 1890, courtesy of Marin County Surveyor)*

Figure 6.15. **Though used for a variety of agricultural purposes over the past 140 years**, wet meadow and depressional wetlands still cover a substantial area of the former Laguna footprint. These wetlands are highly modified, however, and do not provide the same ecological functions as the historical Laguna. *(NAIP 2016)*

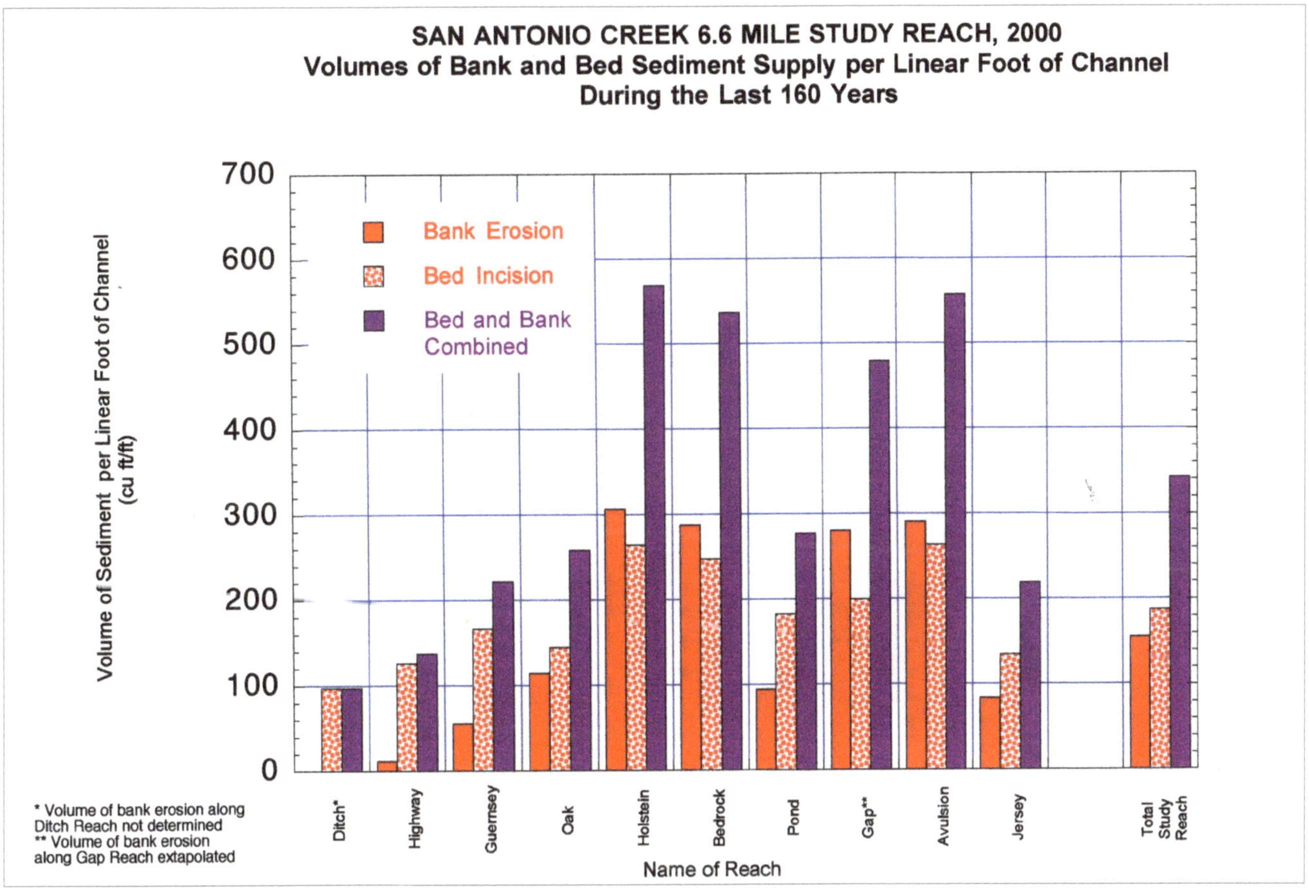

Figure 6.16. **Bank erosion and bed incision**, in part driven by the loss of the Laguna wetland complex and associated changes in peak discharge, has resulted in net increases in sediment supply from the upper and middle reaches of San Antonio Creek over the past 150 years. *(from Collins et al. 2000).*

Figure 6.17. Photographs from 2000 (a) and 2017 (b) showing the progression of bank erosion at a location in San Antonio Creek just south of the confluence of San Antonio Road and I Street. Based on a visual inspection of the photographs, it is estimated that 12 feet or more of bank erosion has occurred in some parts of this reach over this 17 year period. *(a: Photo by Laurel Collins October 24, 2000; b: Photo by Sean Baumgarten August 10, 2017)*

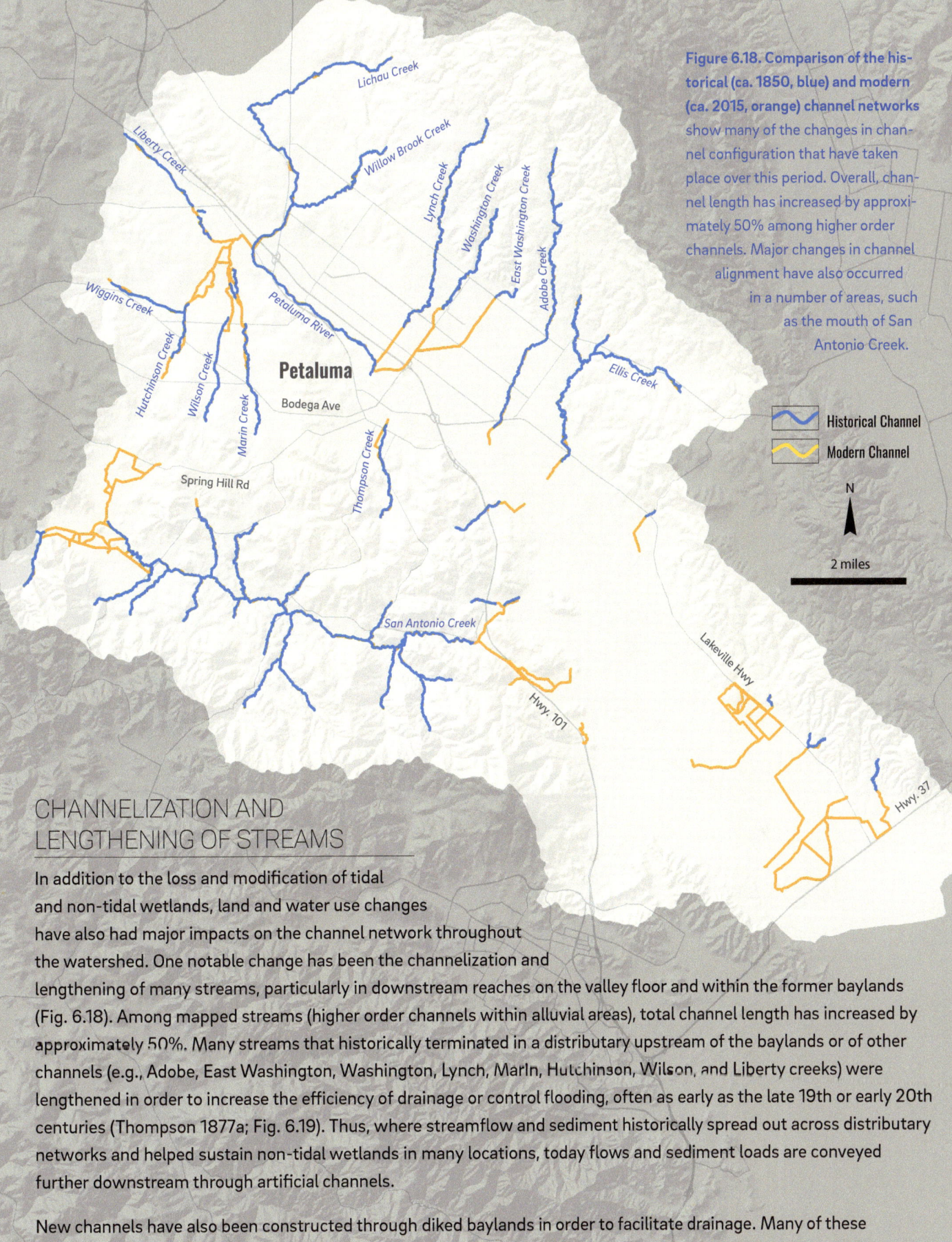

Figure 6.18. Comparison of the historical (ca. 1850, blue) and modern (ca. 2015, orange) channel networks show many of the changes in channel configuration that have taken place over this period. Overall, channel length has increased by approximately 50% among higher order channels. Major changes in channel alignment have also occurred in a number of areas, such as the mouth of San Antonio Creek.

CHANNELIZATION AND LENGTHENING OF STREAMS

In addition to the loss and modification of tidal and non-tidal wetlands, land and water use changes have also had major impacts on the channel network throughout the watershed. One notable change has been the channelization and lengthening of many streams, particularly in downstream reaches on the valley floor and within the former baylands (Fig. 6.18). Among mapped streams (higher order channels within alluvial areas), total channel length has increased by approximately 50%. Many streams that historically terminated in a distributary upstream of the baylands or of other channels (e.g., Adobe, East Washington, Washington, Lynch, Marin, Hutchinson, Wilson, and Liberty creeks) were lengthened in order to increase the efficiency of drainage or control flooding, often as early as the late 19th or early 20th centuries (Thompson 1877a; Fig. 6.19). Thus, where streamflow and sediment historically spread out across distributary networks and helped sustain non-tidal wetlands in many locations, today flows and sediment loads are conveyed further downstream through artificial channels.

New channels have also been constructed through diked baylands in order to facilitate drainage. Many of these channels were excavated in the 1960s and 70s for mosquito abatement (Sanderson et al. 2000). Because these areas historically supported a dense network of tidal channels and sloughs, the constructed channel network represents both a simplification of the former channel network as well as an increase in the length of non-tidal channels.

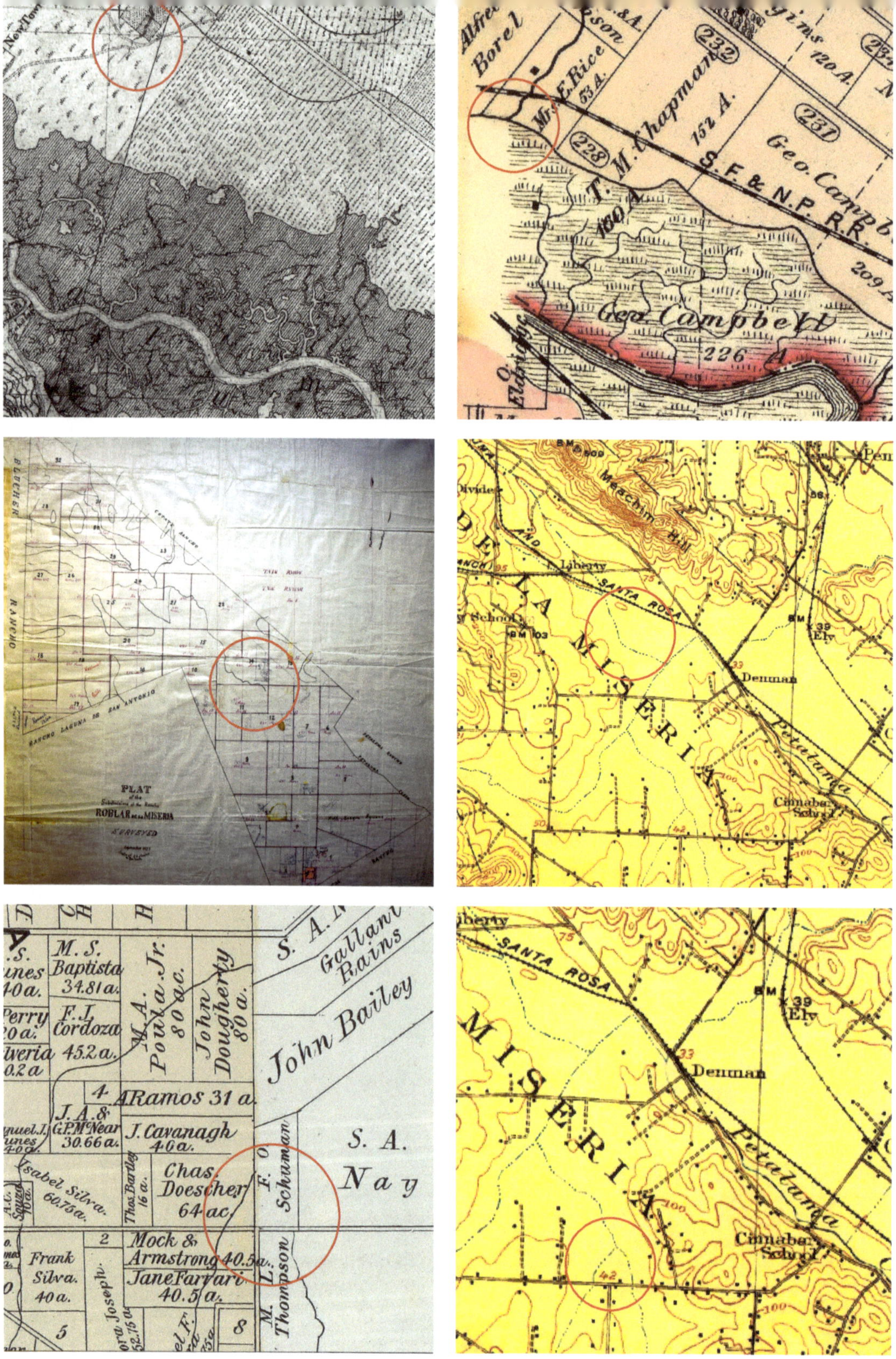

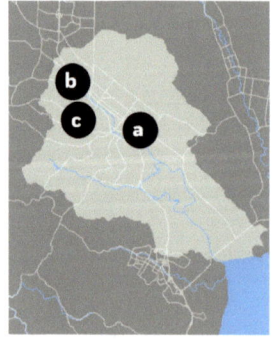

Figure 6.19. Many creeks that historically terminated in a distributary upstream of the tidal marsh or of other channels were lengthened and channelized in order to increase drainage efficiency. (a) By the late 1870s, Adobe Creek had been lengthened to connect with the tidal wetlands just south of Newtown. (b,c) By the early 20th century, Liberty, Marin, and Wilson creeks had been lengthened to connect with the Petaluma River mainstem. *(a (left): Rodgers and Kerr 1860, courtesy of NOAA; a (right): Thompson 1877b, courtesy of David Rumsey Map Collection; b (left): Unknown 1857, courtesy of Curtis & Associates, Inc.; b (right): USGS [1914]1916; c (left): Reynolds and Proctor 1898, courtesy of David Rumsey Map Collection; c (right): USGS [1914]1916)*

CHANGES IN CHANNEL ALIGNMENT

In addition to the overall lengthening of the channel network, the alignment of stream channels has been changed in a number of locations. For instance, the lower portion of Thompson Creek, which historically connected with the Petaluma River near present-day F Street, was straightened and confined to a storm drain sometime during the 1860s or 70s (Fig. 6.20). Portions of many other creeks throughout the watershed, such as Hutchinson, Marin, Washington, and Adobe creeks, and tributaries of San Antonio Creek, have also been ditched and straightened.

The most notable change in channel alignment is at the mouth of San Antonio Creek, which historically entered the tidal marsh to the south of Neil's Island and flowed about 11 km (7 mi) through San Antonio Slough to its confluence with the Petaluma River. In the late 1930s or early 1940s, lower San Antonio Creek was redirected to flow north, into Schultz Slough, which connects with the Petaluma

> *The mouth of Adobe Creek has been changed, and it now empties itself on a mud flat below Newtown.*
> —Petaluma Courier 1879

Figure 6.20. **By 1883, the lower portion of Thompson Creek was channelized** to flow through a storm drain beneath F Street to its confluence with the Petaluma River. *(Left: Rodgers and Kerr 1860, courtesy of NOAA; Right: Sanborn Map and Publishing Co. 1883, courtesy of Earth Sciences & Map Library, UC Berkeley)*

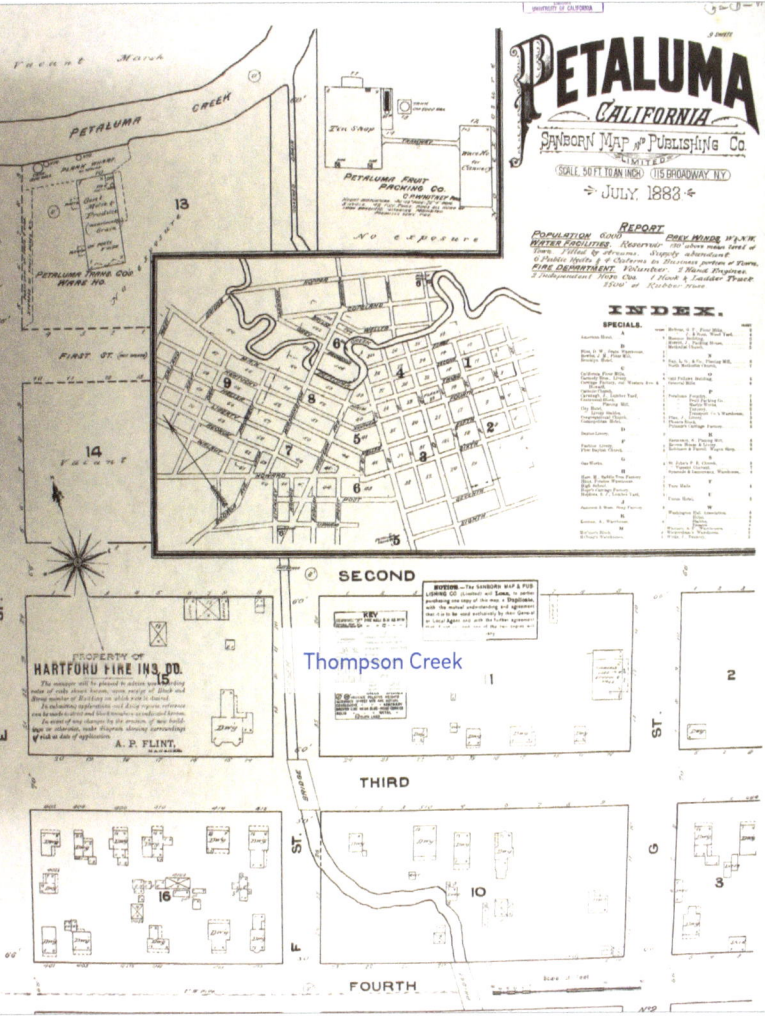

River 8 km (5 mi) further upstream (Fig. 6.21). As a result of this change in channel alignment, the gradient of lower San Antonio Creek has decreased, which, in combination with increased sediment input from erosion in the upper watershed, has led to sediment accumulation, bed aggradation, and an increased rate of levee formation (Prunuske Chatham, Inc. 1998; Collins et al. 2000, Collins n.d.). In contrast, the loss of freshwater input in San Antonio Slough, along with a decrease in tidal prism due to marsh reclamation, has resulted in an overall narrowing of channel width (Collins n.d.).

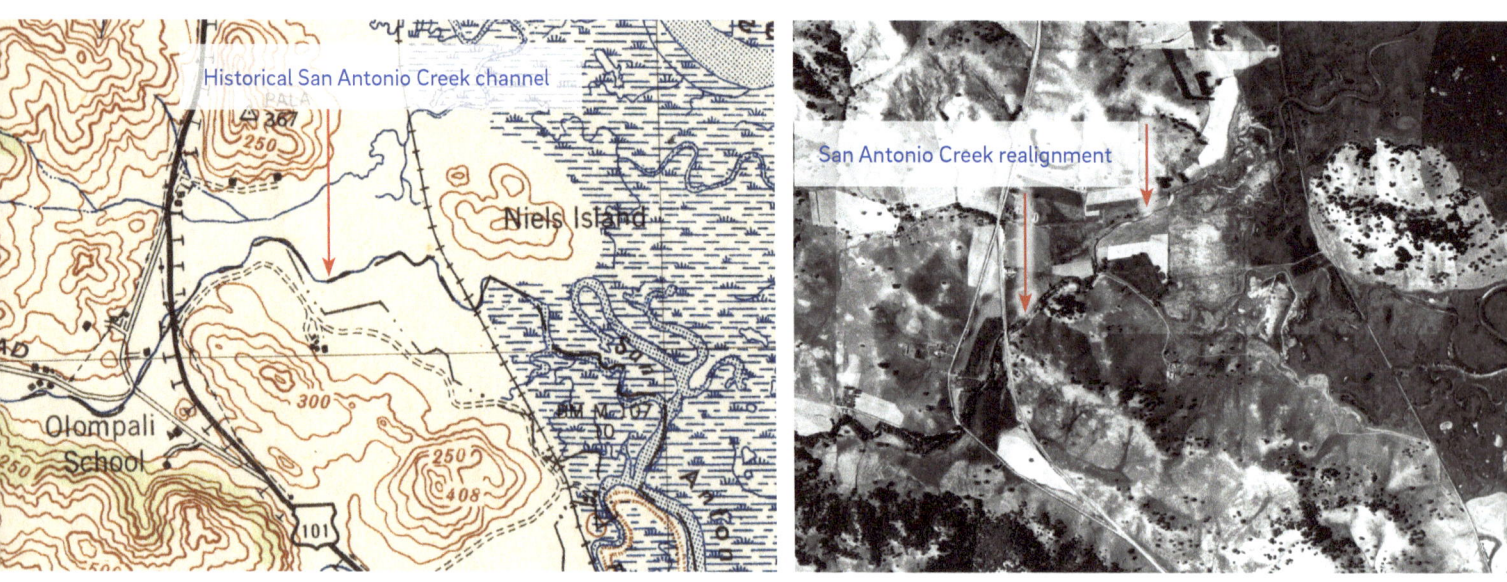

Figure 6.21. The lower reaches of San Antonio Creek, which historically entered the tidal marsh to the south of Neil's Island, were realigned in the late 1930s or early 1940s to flow north into Schultz Slough. The reduced channel gradient created by the new alignment has resulted in sediment accumulation and bed aggradation within lower San Antonio Creek. *(Left: USACE [1937]1942; Right: USDA 1942)*

The depot of the San Francisco and North Pacific Railroad is in East Petaluma, and it grows quite as fast as other portions of the city. The land has been raised by the overflow of the creeks which cross Petaluma valley. These streams formerly spread out over the plain beyond the town, but were gradually confined to a narrow channel, through which this rich tribute from the hills was brought across the plain and spread over the lands of East Petaluma, thereby greatly enhancing their value.

–Thompson 1877a

7. SYNTHESIS AND NEXT STEPS

This study shows through a comparison of historical and contemporary mapping that the Petaluma River watershed has experienced considerable habitat loss over the past 150 years. This includes a 7,340 ha (18,130 ac) decrease in wetland extent (tidal and freshwater) due to land reclamation for agricultural and urban development, and channel realignment and channel lengthening to improve flood conveyance through reclaimed areas (Fig. 7.1). These profound landscape changes have affected ecosystem functions and decreased the overall ecosystem services the watershed once provided. Despite these changes, the large amount of relatively undeveloped land within the watershed also provides extensive opportunities to restore functioning and interconnected wetland habitats (Goals Project 2015). Restoring lost tidal, tidal-terrestrial, and fluvial/upland habitats in this watershed could provide considerable benefits for a wide range of native species such as Ridgway's rail, California red-legged frog, and steelhead, while also providing flood alleviation, groundwater recharge, and stormwater retention and filtration benefits.

Petaluma River. *(photo by Gary H., February 2015, licensed under Creative Commons)*

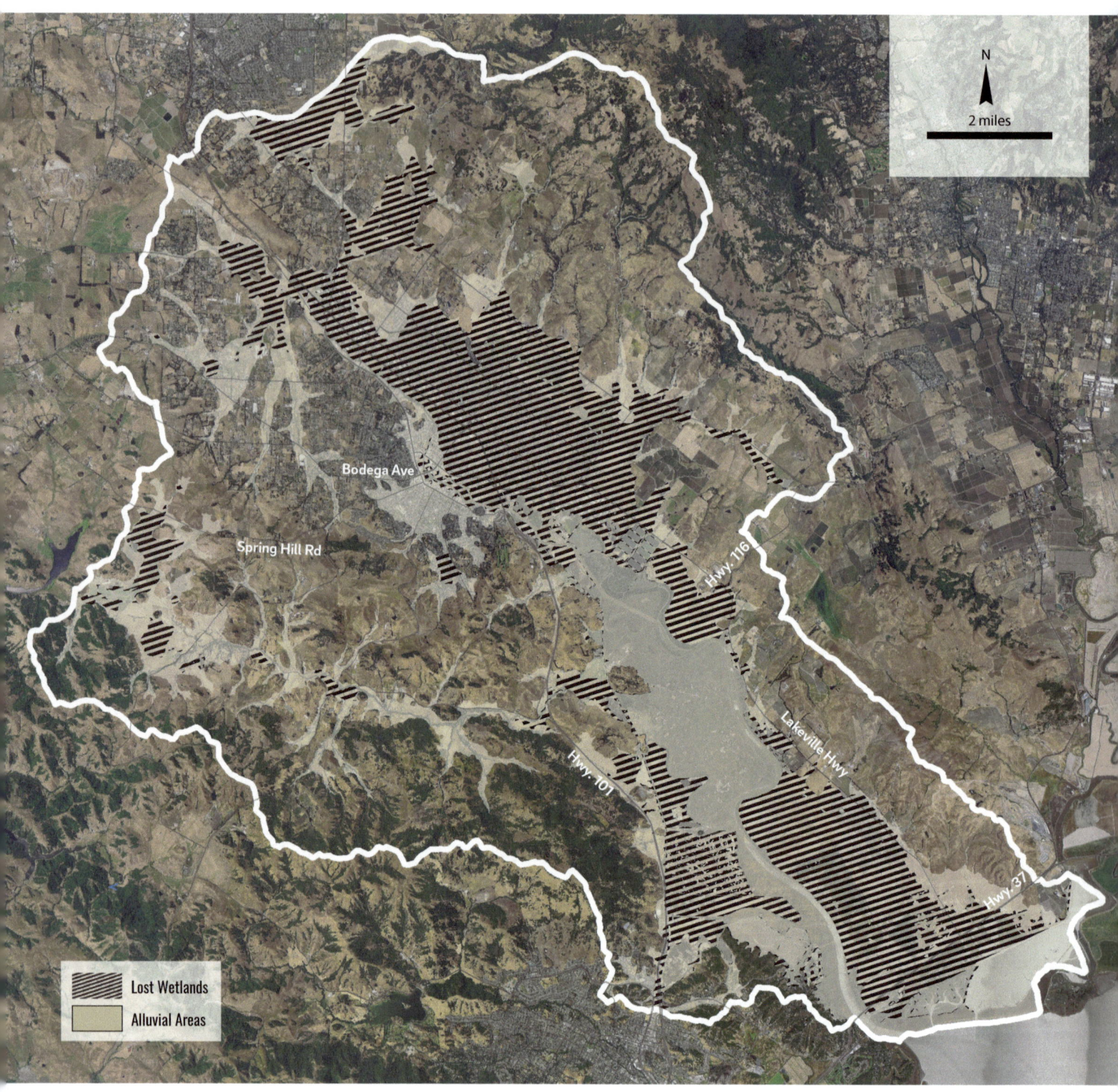

Figure 7.1. The extent of tidal and non-tidal wetlands in alluvial areas has decreased by 7,340 ha (18,130 ac) within the **Petaluma River watershed** over the past two centuries. Wetland loss has greatly reduced habitat for numerous plants and animals, and has also impaired numerous ecosystem services such as floodwater storage and fine sediment retention. *(NAIP 2016)*

The results of the historical hydrology and landscape change analyses were combined with an examination of contemporary physical controls to identify potential opportunity areas throughout the watershed for restoring lost tidal, transitional, and fluvial/upland habitats. These are areas that meet basic criteria that suggest they *could* be suitable for restoration. See page 21 for an overview of the methods and sources used to identify potential restoration opportunity areas.

- **Tidal marsh:** Opportunity areas for tidal marsh restoration along the Petaluma River include those relatively undeveloped areas that were mapped as historical tidal marsh and are now at or below current tidal marsh elevation (i.e., current mean higher high water [MHHW] elevation) (Fig. 7.2). These areas are located mainly around the mouth of the Petaluma River where it enters San Pablo Bay. There are several places near the mouth where it could be possible to reconnect large remnant tidal marsh channels to the Petaluma River as part of an effort to reestablish broad tidal marsh plains and marsh pannes.

- **Tidal-terrestrial transition zone**: Opportunity areas for tidal-terrestrial transition zone restoration include relatively undeveloped areas between the upland edge of existing or potentially restored tidal marshes along the Petaluma River (i.e., MHHW elevation) and 1.8 m (6 ft) above MHHW, which captures the space where marsh could migrate inland over the next several decades using the most recent estimate of extreme sea-level rise by the end of the century (Griggs et al. 2017; see Fig. 7.2). These areas exist primarily within 1 km (0.6 mi) of the historical tidal marsh edge, with some of best opportunity areas located around the mouth of San Antonio Creek and along eastern edge of the marsh around Lakeville.

- **Freshwater wetland:** Opportunity areas for restoring freshwater wetlands (including wet meadows, vernal pools, and valley freshwater marsh) within the watershed include relatively undeveloped areas that: 1) were mapped as historical freshwater wetland; and 2) other areas with poorly drained soils that could potentially support freshwater wetlands (see Fig. 7.2). These areas exist in upland watershed areas and in the tidal-terrestrial transition zone, with the best opportunity areas located along San Antonio Creek (particularly in the headwaters), Lichau Creek, Willow Brook Creek, lower Ellis Creek, and lower Marin Creek, and in the Denman Flat area. Groundwater elevation will be a key determinant of restoration feasibility in many of these areas, and in some cases groundwater depletion may preclude wetland restoration.

- **Riparian forests and wetlands:** Opportunity areas for restoring riparian forests and wetlands along the mainstem Petaluma River and major tributaries include relatively undeveloped areas within the FEMA 100-year floodplain (see Fig. 7.2). These areas are currently inundated during very large flood events, at a minimum, and could therefore have the appropriate hydrology and hydraulics to support native riparian forests similar to what existed historically. As with freshwater wetlands, these areas exist in upland reaches and reaches within the tidal-terrestrial transition zone. Some of the best opportunity areas are located along San Antonio Creek, Petaluma River upstream of downtown Petaluma, lower Lichau Creek, lower Willow Brook Creek, and lower Marin Creek.

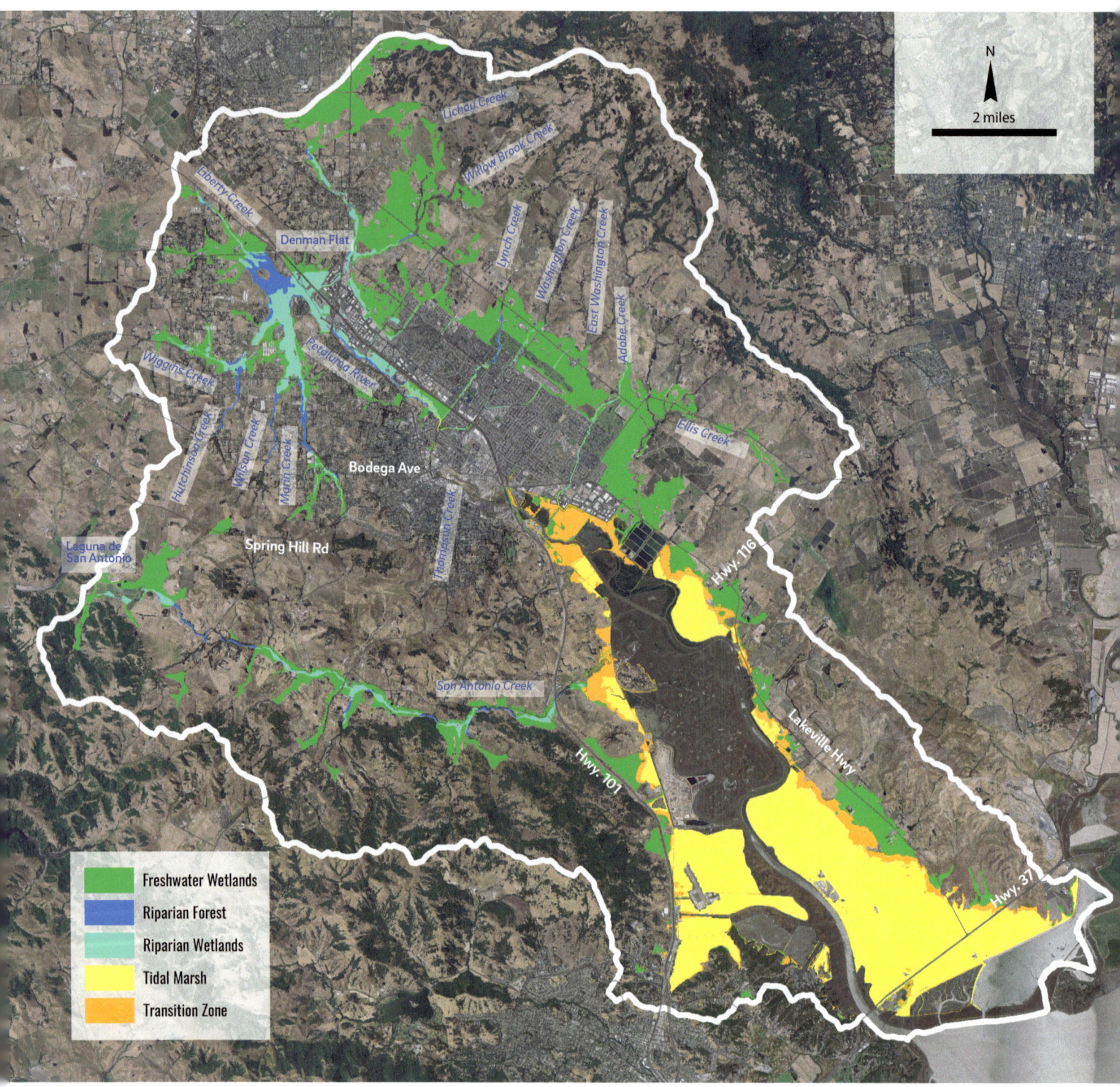

Figure 7.2. Potential opportunities to restore a range of wetland and riparian habitat types, including tidal marsh (yellow), tidal-terrestrial transition zones (orange), freshwater wetlands (green), riparian forests (blue), and riparian wetlands (teal), exist on relatively undeveloped lands throughout the watershed (including lands currently in agricultural use). These potential restoration opportunity areas meet basic physical criteria necessary to support various types of wetland habitats, such as poorly drained soils, periodic flooding, and/or elevations within or near the tidal range. Further analysis, taking into account landowner interest and a range of other physical characteristics, will be needed to determine the feasibility and priority of restoration for particular sites. *(NAIP 2016)*

NEXT STEPS

FEASIBILITY ANALYSIS

In the next phase of this project, a restoration feasibility analysis will be conducted to assess which restoration opportunity areas discussed above have the characteristics needed for viable, self-sustaining habitat restoration projects. Key feasibility considerations include site physical characteristics such as soil quality (i.e., degree of contamination), groundwater elevations, dry season flows, degree of channel incision and overbank flood frequency, and fluvial sediment supply (fine and coarse). Land ownership will be the initial screening factor for determining restoration feasibility. The most promising restoration areas, among those with appropriate physical characteristics, will be those with landowners (either public or private) that have a strong interest in implementing restoration projects. Sonoma RCD will lead the restoration feasibility analysis with support from SFEI-ASC and other local partners. The analysis is expected to be completed by summer 2018.

WATERSHED RESTORATION VISION

Ultimately, a landscape-scale restoration vision is needed to synthesize information about restoration opportunities and feasibility, and to prioritize and guide integrated restoration efforts throughout the watershed. The vision could be modeled on recent landscape visions that SFEI developed for the Novato Creek Baylands (SFEI-ASC 2015b) and Lower Walnut Creek (SFEI-ASC 2016) as part of the EPA-funded Flood Control 2.0 project, which used quantitative information about landscape change and contemporary physical setting to develop specific restoration targets and strategies to enhance desired ecological functions and ecosystem services. This approach is currently being used to develop a restoration plan and vision for the Laguna de Santa Rosa, with funding from the California Department of Fish and Wildlife and Sonoma County Water Agency. Existing studies and plans, including this historical hydrology study, the Draft Petaluma Watershed Enhancement Plan (SRCD 2015), and analyses of restoration opportunities and feasibility (see discussion above) will provide many of the components needed to develop a landscape-scale restoration vision for the Petaluma River watershed.

Petaluma River. *(photo by Greenbelt Alliance, August 2011, licensed under Creative Commons)*

REFERENCES

Adams F, Harding ST, Robertson RD, et al. 1912. *Reports on the irrigation resources of California*. U.S. Department of Agriculture. Sacramento, California.

Alden J. 1860a. Hydrography of Petaluma Creek, California, from Petaluma Point to Lakeville (Sheet No. 724). U.S. Coast Survey. 1:10,000. *Courtesy of NOAA*.

Alden J. 1860b. Hydrography of Petaluma Creek, California, from Lakeville to Petaluma (Sheet No. 725). U.S. Coast Survey. 1:10,000. *Courtesy of NOAA*.

Alley, Bowen & Co., Publishers. 1880. *History of Sonoma County*. Oakland, CA.

Altimira J. 1823. Journal of the expedition verified with the object of examining localities for the founding of the new mission of O.F. San Francisco of Upper California. In *Hutching's California Magazine, no. 50 (August 1860) and no. 51 (September 1860)*. San Francisco: Hutchings & Rosenfield.

Atwater BF, Conrad SG, Dowden JN, et al. 1979. History, landforms, and vegetation of the estuary's tidal marshes. In *San Francisco Bay: The Urbanized Estuary, Investigations into the Natural History of San Francisco Bay and Delta with Reference to the Influence of Man*, ed. San Francisco, California: Pacific Division of the American Association for the Advancement of Science.

Austin H. 1873. Map of Marin County California. San Francisco: A.L. Bancroft. *Courtesy of David Rumsey Map Collection*.

BAOSC (Bay Area Open Space Council). 2017. *The conservation lands network biodiversity portfolio report* [for custom user-defined area]. Accessed September 2017. http://www.openspacecouncil.org

Barnby M, Collins J, Resh V. 1985. Aquatic macroinvertebrate communities of natural and ditched potholes in a San Francisco Bay salt marsh. *Estuarine and Coast Shelf Science* 20:331-347.

Barrett SA. 1908. The geography and dialects of the Miwok Indians. *University of California Publications in American Archeology and Ethnology* 6(2):333-367.

Baumgarten S, Grossinger R, Beller E, et al. 2017. *Historical ecology and landscape change in the Central Laguna de Santa Rosa*. SFEI publication no. 820. SFEI-ASC.

Baye PR, Faber PM, Grewell B. 2000. Tidal marsh plants of the San Francisco Estuary. In *Baylands Ecosystem Species and Community Profiles: Life Histories and Environmental Requirements of Key Plants, Fish and Wildlife. Prepared by the San Francisco Bay Area Wetlands Ecosystem Goals Project*, ed. P.R. Olofson. Oakland, California: San Francisco Bay Regional Water Quality Control Board.

Beagle J, Salomon M, Grossinger R, et al. 2015. *Shifting Shores: marsh expansion and retreat in San Pablo Bay*. SFEI Publication no. 751. SFEI-ASC.

Beller E, Salomon M, Grossinger R. 2010. *Historical vegetation and drainage patterns of Western Santa Clara Valley: a technical memorandum describing landscape ecology in Lower Peninsula, West Valley, and Guadalupe Watershed Management Areas*. SFEI Publication no. 622. SFEI-ASC.

Boivin S. 1998. Down a lazy river. *The Journal of the Sonoma Historical Society* 1:9-12. *Courtesy of Sonoma State University Library.*

Bowers AB. 1866. Map of Sonoma County, California. *Courtesy of David Rumsey Map Collection.*

Bowman JN. 1947. *The area of the mission lands. Courtesy of The Bancroft Library, UC Berkeley.*

Butterworth J. 1997. *Analysis of historic and current hydrologic conditions in the Petaluma River*. Jones & Stokes Associates, Inc. Sacramento, California.

Byrne R, Ingram BL, Starratt S, et al. 1998. Carbon-isotope, diatom, and pollen evidence for late Holocene salinity change in a brackish marsh in the San Francisco Estuary. *Quaternary Research* 55.

California Star. 1848. Cal. Star's Sonoma correspondence. January 8. *Courtesy of California Digital Newspaper Collection.*

Cardwell G. 1958. *Geology and ground water in the Santa Rosa and Petaluma Valley areas, Sonoma County California*. Geological Survey Water-Supply Paper 1427. U.S. Government Printing Office.

City of Petaluma. 2015. *City of Petaluma floodplain management plan.*

CNDDB (California Natural Diversity Database). 2012. California Department of Fish and Game, Biogeographic Data Branch.

CNPS (California Native Plant Society). 2017. Calscape California native plants. October 2017. http://www.rareplants.cnps.org/

Collins JN, Collins LM, Leopold LB, et al. 1986. The influence of mosquito control ditches on the geomorphology of tidal marshes in the San Francisco Bay Area: evolution of salt marsh mosquito habitat. *Proceedings and Papers of the California Mosquito and Vector Control Association* 54.

Collins LM, Collins JN, Leopold LB. 1987. Geomorphic processes of an estuarine tidal marsh: preliminary results and hypotheses. In *International Geomorphology* ed. V. Gardner: John Wiley and Sons, LTD.

Collins JN, Grossinger RM. 2004. *Synthesis of scientific knowledge concerning estuarine landscapes and related habitats of the South Bay Ecosystem. Draft final technical report of the south bay salt pond restoration project*. San Francisco Estuary Institute. Oakland, California.

Collins JN, Resh VH. 1985. Utilization of natural and man-made habitats by the salt marsh song sparrow *Melospiza melodia samuelis* (Baird). *California Fish and Game* 71(1):40-52.

Collins L, Amato P, Morton D. 2000. *SFEI Watershed Science analysis of San Antonio Creek, Sonoma and Marin Counties*. SFEI-ASC.

Collins L. n.d. *Tidal processes along the Petaluma Marsh [PowerPoint presentation]*. Watershed Sciences.

County of Marin. 2015. [North Bay topographic-bathymetric surface model. Compiled from 11 source grid datasets.]

Cornwall C, Moore S, DiPietro D, et al. 2014. *Climate ready Sonoma County: climate hazards and vulnerabilities*. Climate Action 2020: A Regional Program for Sonoma County Communities. Santa Rosa, California.

CSDC (California State Data Center). 2012. Historical census populations of California, counties, and incorporated cities, 1850-2010. http://www.dof.ca.gov/Reports/Demographic_Reports/index.html#reports

Daily Alta California. 1872. Sonoma County: its towns, scenery, climate, natural resources – Petaluma, Santa Rosa, Their Promising Future. October 30. *Courtesy of California Digital Newspaper Collection.*

Dawson A, Salomon M, Whipple A, et al. 2008. *An introduction to the historical ecology of the Sonoma Creek Watershed: a tool for developing an action plan for the critical coastal areas program*. San Francisco Estuary Institute and Sonoma Ecology Center.

Dawson A, Sloop C. 2010. *Laguna de Santa Rosa historical hydrology project headwaters pilot study*. Sonoma Ecology Center and Laguna de Santa Rosa Foundation.

Dickerson RE. 1922. Tertiary and quaternary history of the Petaluma, Point Reyes and Santa Rosa quadrangles. *Proceedings of the California Academy of Sciences* 11(19):527-601. San Francisco, California.

Dixon J. 1908. *Petaluma, Sonoma Co., Calif. [Field Notes v583]*. Courtesy of Museum of Vertebrate Zoology, UC Berkeley.

Dixon J. 1909. A new harvest mouse from Petaluma, California. *University of California Publications in Zoology* 5(4): 271-273. Berkeley: The University Press. *Courtesy of Google Books*.

Dolcini Family. 1984, 2006, 2007. Dolcini family interviews, Marin County ranchers since 1856. Dewey Livingston, Jow NcNeil, Carla Ehat, and Marilyn Geary, editors. Marin Resource Conservation District.

DWR (Department of Water Resources). 1982. *Evaluation of ground water resources Sonoma County, volume 3: Petaluma Valley*. Department of Water Resources.

Engelhardt Z. 1897. *The Franciscans in California*. Harbor Springs, Michigan: Holy Childhood Indian School.

Feathers DA. 1935. VertNet, record for *Sorex vagrans sonomae* from "3 miles south of Petaluma." American Museum of Natural History (AMNH).

FEMA (Federal Emergency Management Administration). 2016. National flood hazard layer (NFHL). [Sonoma County, 06041C.]. U.S. Department of Homeland Security. Accessed July 2017. https://fema.maps.arcgis.com/home/index.html

FEMA (Federal Emergency Management Administration). 2017. National flood hazard layer (NFHL). [Sonoma County, 06097C.]. U.S. Department of Homeland Security. Accessed July 2017. https://fema.maps.arcgis.com/home/index.html

Fisher G. 1852. *United States vs. Mariano G. Vallejo*. Deposition of G. Fisher. Land Case File 321 N.D. U.S. District Court, Northern District. *Courtesy of The Bancroft Library, UC Berkeley*.

Foin T, Garcia E, Gill R, et al. 1997. Recovery strategies for the California clapper rail (*Rallus longirostris obsoletus*) in the heavily urbanized San Francisco estuarine ecosystem. *Urban Planning* 38:229-243.

Ghodrati F, Lunde K. 2017. *Petaluma River bacteria and nutrients TMDL: stakeholder meeting*. California Water Boards.

Goals Project. 2015. *The baylands and climate change: what we can do. Baylands Ecosystem Habitat Goals Science Update 2015*. California State Coastal Conservancy. Oakland, California.

Goudie A, editor. 2004. *Encyclopedia of geomorphology*. London, England: Routledge.

Gonzalez S. 1931. Map of Petaluma and vicinity showing right-of-way line required for the widening of Petaluma Creek. *Courtesy of Curtis & Associates, Inc.*

Griggs G, Cayan D, Tebldi C, et al. 2017. *Rising seas in California: an update on sea-level rise science*. California Ocean Science Trust and California Ocean Protection Council.

Grossinger RM. 2005. Documenting Local Landscape Change: The Bay Area Historical Ecology Project. In *The Historical Ecology Handbook: A Restorationist's Guide to Reference Ecosystems*, ed. Dave Egan and Evelyn A Howell. Washington D.C.: Island Press.

Grossinger RM, Askevold RA, Collins JN. 2005. *T-sheet user guide: application of the historical U.S. Coast Survey maps to environmental management in the San Francisco Bay Area. A technical report of SFEI's Historical Ecology and Wetlands Programs.* SFEI publication no. 427. San Francisco Estuary Institute. Oakland, California.

Grossinger RM et al. 2006. *Coyote Creek watershed historical ecology study: Historical condition, landscape change, and restoration potential in the Eastern Santa Clara Valley, California.* SFEI publication no. 426. SFEI-ASC.

Grossinger RM, Striplen CJ, Askevold RA, et al. 2007. Historical landscape ecology of an urbanized California valley: wetlands and woodlands in the Santa Clara Valley. *Landscape Ecology* 22:103-120.

Grossinger RM. 2012. *Napa Valley historical ecology atlas: exploring a landscape of transformation and resilience*. Berkeley, California: University of California Press.

Hanley N, Ready R, Colombo S, et al. 2009. The impacts of knowledge of the past on preferences for future landscape change. *Journal of Environmental Management* 90:1404-1412.

Healdsburg Enterprise. 1878. News from our neighbors. March 5. *Courtesy of California Digital Newspaper Collection.*

Healdsburg Tribune. 1925. River and creeks overflow when 4.44 inches of rain falls here in 24 hours; Rail service abandoned. February 11. *Courtesy of California Digital Newspaper Collection.*

Heig A. 1982. *History of Petaluma: a California river town*. Petaluma, California: Scottwall Associates.

Heuer WH. 1917. Survey of Petaluma Creek, Cal. In *Letter from the Secretary of War transmitting, with a letter from the Chief of Engineers, reports on preliminary examination and survey of Petaluma Creek, Cal. H. Doc. 849, 65-2.*

Higgs E. 2012. History, novelty, and virtue in ecological restoration. In *Ethical adaptation to climate change: human virtues of the future*, ed. Allen Thompson and Jeremy Bendik-Keymer. Cambridge, Massachusetts: MIT Press.

Holmes LC, Nelson JW. 1914. Soil Map: California Reconnaissance Survey—San Francisco Bay Sheet. USDA (United States Department of Agriculture).

Holmes LC, Nelson JW. 1917. Reconnoissance soil survey of the San Francisco Bay region, California. USDA (United States Department of Agriculture).

Holway RS. 1907. Physiographic changes bearing on the faunal relationships of the Russian and Sacramento rivers, California. *Science* 26(664):382-383.

Holway RS. 1914. Physiographically Unfinished Entrances to San Francisco Bay. *University of California Publications in Geography* 1(3).

Irelan W. 1890. *Tenth annual report of the State Mineralogist for the year ending December 1, 1890*. California State Mining Bureau. Sacramento, California.

Irelan W. 1893. *Eleventh report of the State Mineralogist, two years ending September 15, 1892*. California State Mining Bureau.

Jackson S, Hobbs R. 2009. Ecological restoration in the light of ecological history. *Science* 325:567-569.

Jackson T. 1927. Survey of Petaluma Creek, Calif. In *Letter from the Secretary of War transmitting report from the Chief of Engineers on preliminary examination and survey of Petaluma Creek, Calif. H. Doc. 183, 70-1. Correspondence concerning the Petaluma River and dredging projects, 1890-1926.*, ed. *Courtesy of Sonoma County Library Petaluma History Room.*

Jepson WL. 1928. The botanical explorers of California II. *Madroño* 1(11):175-177. California Botanical Society.

Kelley, RB. 2017. *Plagiobothrys mollis* var. *vestitus*, in *Jepson Flora Project* (eds.). Jepson eFlora.

Kennedy C, Grow P, Harrison T. 2012. *Upper Petaluma River watershed flood control project scoping study*. RMC Water and Environment.

Kerr D. 1860. Map of part of Petaluma Creek California. Register No. 817. United States Coast Survey (USCS). *Courtesy of National Oceanic and Atmospheric Administration.*

Koch L. 1940. Consortium of California Herbaria, record for *Ruppia maritima* from "Along Hwy. 101, 2 mi. S of Sonoma County line… In water of saline flat." Humboldt State University Herbarium (HSC).

Lamb CC. 1927. *Petaluma, Sonoma Co., Calif. [Field Notes v1420 part 2]. Courtesy of Museum of Vertebrate Zoology, UC Berkeley.*

Langenheim V, Graymer R, Jachens R, et al. 2010. Geophysical framework of the northern San Francisco Bay region, California. *Geosphere* 6(5):594-620.

Lawson JS. 1886-7. *U.S. Coast and Geodetic Survey Descriptive Report*. U.S. Coast and Geodetic Survey. *Courtesy of NOAA.*

Lawson JS, Welker PA. 1887. Re-survey of San Pablo Bay. Register number 1827. United States Coast and Geodetic Survey (USCGS). *Courtesy of NOAA.*

Lee A. 1901. *Some sportsmen's clubs of California where sport is perennial*. Outing: An Illustrated Magazine of Sport and Travel Adventure & Country Life. New York, New York: The Outing Publishing Company.

Leidy RA, Becker GS, Harvey BN. 2005. *Historical distribution and current status of steelhead/rainbow trout (Oncorhynchus mykiss) in streams of the San Francisco Estuary, California*. CEMAR (Center for Ecosystem Management and Restoration).

Leidy RA. 2007. *Ecology, assemblage structure, distribution, and status of fishes in streams and tributaries to the San Francisco Estuary, California*. U.S. Environmental Protection Agency.

Leopold LB, Collins JN, Collins LM. 1993. Hydrology of some tidal channels in estuarine marshland near San Francisco. *Catena* 20:469-493.

Lewis Publishing Company. 1891. *A memorial and biographical history of Northern California, illustrated.* Chicago.

Lightfoot KG, Parrish O. 2009. *California Indians and their environment: an introduction*. Berkeley, California: University of California Press.

Loring FR. 1852. *Township lines, south and west of Sonoma*. Book R-210. U.S. Department of the Interior, Bureau of Land Management. *Courtesy of BLM.*

Lynch J, Lynch W, Lynch C. 1872. *Certificate of incorporation of the Sonoma Mountain Irrigation Company*. BANC MSS 73/106 c. *Courtesy of The Bancroft Library, UC Berkeley.*

Mailliard J. 1879. VertNet, record for *Agelaius tricolor* from "Petaluma." California Academy of Sciences (CAS).

Malamud-Roam F, Goman MF. 2012. Historical formation. In *Ecology, Conservation, and Restoration of Tidal Marshes: The San Francisco Estuary*, ed. Arnas Palaima. Berkeley, California: University of California Press.

Marin Journal. 1878. Real estate. June 27. *Courtesy of California Digital Newspaper Collection.*

Marin Journal. 1901. Chileno Valley items. February 14. *Courtesy of California Digital Newspaper Collection.*

Marin Journal. 1915. State highway. April 8. *Courtesy of California Digital Newspaper Collection*.

Martin HB. 1862. *Survey No. 23-24*. Sonoma County Ca. official surveys vol. I 1850-1873.

Matthewson RC. 1859a. *Field notes of the final survey of the Rancho Olompali, Camilo Ygnita, confirmee*. Book G-12. U.S. Department of the Interior, Bureau of Land Management. *Courtesy of BLM*.

Matthewson RC. 1859b. Plat of the Rancho Olompali. San Francisco, California: General Land Office. 4 chains: 1 inch. *Courtesy of BLM*.

Mendell GH. 1883. Improvement of harbors of Oakland and Wilmington; of Sacramento and San Joaquin Rivers; of Petaluma Creek, and of Humboldt Bay and Harbor, California. In *Annual report of the Chief of Engineers, United States Army, to the Secretary of War, for the year 1883. In three parts, Part III*. H. Doc. Ex. Doc. 1, pt. 2, vol. II, 48-1.

Menefee CA. 1873. *A Brief History of Napa, Sonoma, Lake and Mendocino Counties prior to 1873*. Fresno, California: California History Books.

Milliken RA. 2009. *Ethnohistory and ethnogeography of the Coast Miwok and their neighbors, 1783-1840*. A technical report prepared for the National Park Service, Golden Gate National Resources Area, Cultural Resources and Museum Management Division, Archaeological / Historical Consultants. Oakland, California.

Morning Call. 1890. Fishing and dredging. October 27. *Courtesy of Newspapers.com*.

Munro-Fraser JP. 1880. *History of Sonoma County including its geology, topography, mountains, valleys and streams*. Oakland, California: Alley, Bowen & Co.

NAIP (National Agricultural Imagery Program). 2016. [Natural color aerial photos of Sonoma and Marin counties.] Ground resolution: 60 cm. National Agricultural Imagery Program (NAIP). U.S. Department of Agriculture (USDA), Washington, D.C.

National Audubon Society. Guide to North American Birds.2017. Accessed September 2017. http://www.audubon.org/bird-guide

NatureServe. 2017. NatureServe Explorer: An online encyclopedia of life [web application]. Version 7.1. NatureServe, Arlington, Virginia. Accessed July 2017. http://explorer.natureserve.org

Newberry JS. 1857. *Zoological report: explorations and surveys for a railroad route from the Mississippur River to the Pacific Ocean*. War Department. Washington D.C.

NOAA (National Oceanic and Atmospheric Administration). 2012a. NOAA Coastal Services Center sea level rise data: 1-6 ft sea level rise inundation extent. Department of Commerce. Charleston, S.C. Accessed July 2017. http://www.csc.noaa.gov/slr

NOAA (National Oceanic and Atmospheric Administration). 2012b. NOAA Coastal Services Center sea level rise data: current mean higher high water inundation extent. Department of Commerce. Charleston, S.C. Accessed July 2017. http://www.csc.noaa.gov/slr

NRCS (Natural Resources Conservation Service). 2011. Soil survey geographic (SSURGO) database for Sonoma County, California, ca097.: U.S. Department of Agriculture, Natural Resources Conservation Service.

NRCS (Natural Resources Conservation Service). 2013. Soil survey geographic (SSURGO) database for Marin County, California, ca041.: U.S. Department of Agriculture, Natural Resources Conservation Service.

Panoramics. 1860. *A trip to the California geysers*. Hutching's California Magazine. San Francisco, California: Hutchings & Rosenfield.

Parker GF. 1868. *Gelo Freeman Parker diaries and letterbook*. BANC MSS 72/168 c. *Courtesy of The Bancroft Library, UC Berkeley*.

Parker GF. 1869. *Gelo Freeman Parker diaries and letterbook*. BANC MSS 72/168 c. *Courtesy of The Bancroft Library, UC Berkeley.*

Parker GF. 1870. *Gelo Freeman Parker diaries and letterbook*. BANC MSS 72/168 c. *Courtesy of The Bancroft Library, UC Berkeley.*

Pemberton JR. 1902. VertNet, record for *Melospiza lincolnii gracilis* from "Petaluma." Museum of Vertebrate Zoology, UC Berkeley (MVZ).

Petaluma Courier. 1879. Jettings. *January 29. Courtesy of Newspapers.com.*

Petaluma Courier. 1887a. Courierlets. May 18. *Courtesy of Newspapers.com.*

Petaluma Courier. 1887b. Courierlets. November 2. *Courtesy of Newspapers.com.*

Petaluma Weekly Argus. 1877a. June 22. *Courtesy of Newspapers.com.*

Petaluma Weekly Argus. 1877b. Plenty of water. July 6. *Courtesy of Newspapers.com.*

Petaluma Weekly Argus. 1881a. Local brevities. December 4. *Courtesy of Newspapers.com.*

Petaluma Weekly Argus. 1881b. The storm. December 4. *Courtesy of Newspapers.com.*

Petaluma Weekly Argus. 1886. Old paths. November 6. *Courtesy of Newspapers.com.*

Piper CV. 1920. A study of allocarya. *Contributions from the United States National Herbarium* 22(2):79-113. Department of Botany, Smithsonian Institution.

Press Democrat. 1899. It is a great rain: the precipitation causes the creeks to rise, lower portion of Petaluma covered by the flood and the river over its banks. March 25. *Courtesy of California Digital Newspaper Collection.*

Price JR, Nurse MA. 1896. *Two plans for protecting the City of Petaluma from overflow water, and for improving navigation of Petaluma Creek.* California Department of Public Works. Sacramento, California.

Prunuske Chatham I. 1998. Appendix E: erosion and sedimentation in the Petaluma River Watershed. In *Petaluma River Watershed Enhancement Plan*, ed. Southern Sonoma County Resource Conservation District. Occidental, California.

Reynolds and Proctor. 1898. T. 5 N., R. 8 W. in Illustrated atlas of Sonoma County, California. Compiled and published from personal examinations, official records and actual surveys. Santa Rosa, CA: Reynolds & Proctor. *Courtesy of David Rumsey Map Collection.*

Rhemtulla JM, Mladenoff DJ. 2007. Why history matters in landscape ecology. *Landscape ecology* 22(1-3).

Richardson WA. 1853. *Charles White vs. United States.* Deposition of W.A. Richardson. Land Case File 346 N.D. *Courtesy of The Bancroft Library, UC Berkeley.*

Rodgers AF. 1854. San Francisco Bay, California. Register number 472. United States Coast Survey (USCS). *Courtesy of NOAA.*

Rodgers AF, Kerr D. 1860. Map of part of Petaluma Creek California Plane Table Sheet No. 6. Register number 818. United States Coast Survey (USCS). *Courtesy of NOAA.*

Rodgers TL. 1940. VertNet, record for *Rana draytonii* from "3.5 mi ENE Petaluma." Museum of Vertebrate Zoology, UC Berkeley (MVZ).

Rodgers TL, Stirton RA. 1940. VertNet, record for *Rana boylii* from "3.5 mi ENE Petaluma." Museum of Vertebrate Zoology, UC Berkeley (MVZ).

Roop W, Flynn K. 2007. *A few historical facts about Petaluma and its River.* Archaeological Resource Service and ABACUS Archaeological Associates.

Sacramento Daily Union. 1862. News of the morning, trout fishing. April 14. *Courtesy of California Digital Newspaper Collection*.

Safford H, Wiens JA, Hayward GD. 2012. The growing importance of the past in managing ecosystems of the future. In *Historical environmental variation in conservation and natural resource management*, ed. Gregory Hayward John Wiens, Hugh Safford, and Catherine Giffen, 319-327: John Wiley & Sons.

Salomon et al. 2008. The historical ecology of Mill Creek. San Francisco Estuary Institute, Richmond, CA. http://www.nbwatershed.org/millercreek/

Salomon M, Baumgarten SA, Dusterhoff, SR, Beller EE, Grossinger RM, Askevold RA. 2015. *Novato Creek Baylands historical ecology study.* A Report of SFEI-ASC's Resilient Landscapes Program, Publication #740, San Francisco Estuary Institute-Aquatic Science Center, Richmond, CA.

Samuels E. 1856a. VertNet, record for *Icterus galbula bullockii* from "Petaluma." National Museum of Natural History, Smithsonian Institution (USNM).

Samuels E. 1856b. VertNet, record for *Passerina amoena* from "Petaluma." National Museum of Natural History, Smithsonian Institution (USNM).

Samuels E. 1856c. VertNet, record for *Picoides nuttallii* from "Petaluma." National Museum of Natural History, Smithsonian Institution (USNM).

Samuels E. 1856d. VertNet, record for *Sorex ornatus ornatus* from "Petaluma." National Museum of Natural History, Smithsonian Institution (USNM).

San Francisco Call. 1890. Fishing and dredging. October 27. *Courtesy of California Digital Newspaper Collection*.

San Francisco Call. 1900. Negotiating for tract of sugar-beet land. December 17. *Courtesy of Newspapers.com*.

San Francisco Call. 1913. 1,000 acre Petaluma Ranch sold for cash. February 15. *Courtesy of California Digital Newspaper Collection.*

Sanborn Map and Publishing Co. 1883. Petaluma, California. *Courtesy of Earth Sciences & Map Library, UC Berkeley.*

Sanderson EW, Foin TC, Ustin SL. 2001. A simple empirical model of salt marsh plant spatial distributions with respect to a tidal channel network. *Ecological Modelling* 139:293-307.

Sanderson EW, Ustin SL, Foin TC. 2000. The influence of tidal channels on the distribution of salt marsh plant species. *Plant Ecology* 146:29-41.

Schoellhamer D. 2003. Sediment transport explains contaminant distribution: Petaluma River. In *Pulse of the Bay: Monitoring & Managing Contamination in the San Francisco Estuary*, ed. SFEI-ASC.

Schulz J. 1927. Preliminary examination of Petaluma Creek, Calif. In *Letter from the Secretary of War transmitting report from the Chief of Engineers on preliminary examination and survey of Petaluma Creek, Calif. H. Doc. 183, 70-1. Correspondence concerning the Petaluma River and dredging projects, 1890-1926.*, ed. *Courtesy of Sonoma County Library Petaluma History Room.*

SCWA (Sonoma County Water Agency). 1986. *Petaluma River Watershed drainage plan.*

SFBRWQCB (San Francisco Bay Regional Water Quality Control Board). 2016. 2016 California 303(d) List of Water Quality Limited Segments.

SFEI (San Francisco Estuary Institute). 2017. *Bay Area EcoAtlas: Geographic information system of wetland habitats past and present.*

SFEI-ASC (San Francisco Estuary Institute-Aquatic Science Center). 2015a. *Bay Area Aquatic Resources Inventory (BAARI) version 2 GIS*.

SFEI-ASC (San Francisco Estuary Institute-Aquatic Science Center). 2015b. *Novato Creek Baylands vision: integrating ecological functions and flood protection within a climate-resilient landscape*. A SFEI-ASC Resilient Landscape Program report developed in cooperation with the Flood Control 2.0 project Regional Science Advisors and Marin County Department of Public Works, Publication #764, San Francisco Estuary Institute-Aquatic Science Center, Richmond, CA.

SFEI-ASC (San Francisco Estuary Institute-Aquatic Science Center). 2016. *Resilient landscape vision for Lower Walnut Creek: baseline information & management strategies*. A SFEI-ASC Resilient Landscape Program report developed in cooperation with the Flood Control 2.0 Regional Science Advisors and Contra Costa County Flood Control and Water Conservation District, Publication #782, San Francisco Estuary Institute-Aquatic Science Center, Richmond, CA.

SFEI-ASC (San Francisco Estuary Institute-Aquatic Science Center). 2017. *Changing channels: regional information for developing multi-benefit flood control channels at the Bay interface*. A SFEI-ASC Resilient Landscape Program report developed in cooperation with the Flood Control 2.0 Regional Science Advisors, Publication #801, San Francisco Estuary Institute-Aquatic Science Center, Richmond, CA.

SFEP (San Francisco Estuary Partnership). 2014. *Marin Audubon Society's Bahia Wetland Restoration Project 2009-2014*.

Shellhammer H. 2012. Small mammals. In *Ecology, Conservation, and Restoration of Tidal Marshes: The San Francisco Estuary*, ed. Arnas Palaima. Berkeley, California: University of California Press.

Silliman S. 2004. *Lost laborers in colonial California: Native Americans and the archaeology of Rancho Petaluma*. Tucson, AZ: The University of Arizona Press.

Smith DT. 1906. The geology of the upper region of the main Walker River, Nevada. *University of California Publications in Geography* 4(1).

Sonoma County Journal. 1855. Sheriff's sale. December 22. *Courtesy of Newspapers.com*.

Sonoma County Journal. 1862. Works well. February 14. *Courtesy of Newspapers.com*.

Sonoma Democrat. 1871. The storm throughout the state. December 30. *Courtesy of California Digital Newspaper Collection*.

Sonoma Veg Map. 2017. Sonoma County fine scale vegetation and habitat map. Sonoma County Water Agency, Sonoma County Agricultural Preservation and Open Space District, Sonoma County Vegetation Mapping and LiDAR Program. Accessed July 2017. http://sonomavegmap.org/

SRCD (Sonoma Resource Conservation District). 2015. *Draft Petaluma Watershed enhancement plan: an owner's manual for the residents & landowners of the Petaluma Watershed*.

SSCRCD (Southern Sonoma County Resource Conservation District). 2008. San Antonio Creek Watershed Plan.

Stanford et al. 2013. *Alameda Creek Watershed Historical Ecology Study*. SFEI Publication no. 679. SFEI-ASC.

Stillinger RA. 1982. The Petaluma River: the need for regional historical research to identify archaeological resources. Master of Arts, Cultural Resource Management, Sonoma State University.

Stindt F, Dunscomb G. 1964. The Northwestern Pacific Railroad Redwood Empire route. Redwood City and Modesto, California: Fred A. Stindt and Guy L. Dunscomb.

Stone JR, Curme MA. 1975. The Petaluma River from 1914 to 1950. [Interview with John R. Stone, October 30, 1975.] California State College, Sonoma, Oral History Program. *Courtesy of Sonoma State University Special Collections*.

Swainson OW. 1922. *Descriptive report topographic sheet no. 4015 (mouth to San Antonio Creek), 4016 (San Antonio Cr. to Petaluma)*. U.S. Coast and Geodetic Survey. *Courtesy of NOAA.*

Swainson OW, Overshiner WH, Mower LM. 1922. Register number 4015. United States Coast and Geodetic Survey (USCGS). *Courtesy of NOAA.*

Swetnam TW, Allen CD, Betancourt JL. 1999. Applied historical ecology: using the past to manage for the future. *Ecological Applications* 9(4):1189-1206.

Taylor B. 1862. *At home and abroad: a sketch-book of life, scenery and men*. New York, New York: G.P. Putnam.

Thompson PR. 1857a. *Field notes of the final survey of the Rancho Laguna de San Antonio, B. Bojorguez, confirmee*. Book G-14. U.S. Department of the Interior, Bureau of Land Management. *Courtesy of BLM.*

Thompson PR. 1857b. *Field notes of the final survey of the Rancho Roblar de la Miseria, Daniel Wright, et al, confirmee*. Book G-9. U.S. Department of the Interior, Bureau of Land Management. *Courtesy of BLM.*

Thompson PR. 1857c. Plat of the Rancho Roblar de la Miseria finally conformed [sic] to Daniel Wright et al. General Land Office. 4 chains: 1 inch. *Courtesy of BLM.*

Thompson GH. 1864. *Field notes of the subdivisions and meanders in Township 4 North Range 8 West*. Book R-179. U.S. Department of the Interior, Bureau of Land Management. *Courtesy of BLM.*

Thompson GH. 1871. Map of a tract of marsh-land in Sonoma County Cal. belonging to the San Pablo Land Company. *Courtesy of Curtis & Associates, Inc.*

Thompson RA. 1877a. *Historical and descriptive sketch of Sonoma County, California.* Philadelphia, Pennsylvania: LH Everts & Co.

Thompson TH. 1877b. *Historical atlas of Sonoma County, California*. Oakland, California: Thos. H. Thompson & Co. *Courtesy of David Rumsey Map Collection.*

UCANR (University of California Division of Agriculture and Natural Resources). 2017. California Fish Species. September 2017. http://calfish.ucdavis.edu/species/

United States Army War Department. 1933. *Report of the Chief of Engineers U.S. Army 1933*. War Department. Washington D.C.

Unknown. 1857. Plat of the subdivision of the Rancho Roblar de la Miseria. *Courtesy of Curtis & Associates, Inc.*

Unknown. 1875. Sonoma and Marin railroad survey map. William Hammond Hall Papers, item number 5290-15. *Courtesy of California State Archives.*

Unknown. 1890. Map of the Walker Villa Tract in Marin County, Cal., 6 miles southwest of Petaluma. *Courtesy of Marin County Recorder.*

USACE (U.S. Army Corps of Engineers). [1937]1942. Petaluma Quadrangle, California: 15 minute series (topographic). 1:62,500.

USACE (U.S. Army Corps of Engineers). 1974. *Draft environmental impact report: maintenance dredging, Petaluma River, Sonoma and Marin counties, California.*

USCS (U.S. Coast Survey). 1861. Petaluma and Napa Creeks, California. *Courtesy of NOAA.*

USCGS (U.S. Coast and Geodetic Survey). 1941-2. Planimetric Map, T-5933, California, San Pablo Bay, Petaluma Creek, Novato and vicinity. Register number 5933. United States Coast and Geodetic Survey (USCGS). *Courtesy of NOAA.*

USCGS (U.S. Coast and Geodetic Survey). 1952. Comparitive [sic] cross-sections of Petaluma Creek, vicinity of Black Point. Hydrography from U.S.C. and G.S. United States Coast and Geodetic Survey (USCGS). *Courtesy of State Lands Commission.*

USDA (U.S. Department of Agriculture). 1942. [Aerial photos of Marin and Sonoma Counties] Flight COF-1942. U.S. Department of Agriculture, Soil Conservation Science.

USDA (U.S. Department of Agriculture). 2016. National Agricultural Statistics Service Cropland Data Layer. USDA-NASS, Washington, DC. Accessed July, 2017. https://nassgeodata.gmu.edu/CropScape/

USDC (U.S. District Court, Northern District). ca. 1844. Rancho de Bartolome Bojorce: [Rancho Laguna de San Antonio, Calif.] Land Case Map D-135. *Courtesy of The Bancroft Library, UC Berkeley.*

USDC (U.S. District Court, Northern District). 1845. Plan del Rancho de la Roblar de la Miseria [Calif.]. Land Case Map D-715. *Courtesy of The Bancroft Library, UC Berkeley.*

USDC (U.S. District Court, Northern District). 1852a. Map of the land of Petaluma: [Sonoma Co., Calif.]. Land Case Map D-647. *Courtesy of The Bancroft Library, UC Berkeley.*

USDC (U.S. District Court, Northern District). 1852b. Ranch of Roblar de Meseria: [Sonoma Co., Calif.]. Land Case Map E-64. *Courtesy of The Bancroft Library, UC Berkeley.*

USDC (U.S. District Court, Northern District). ca. 1856. [Diseño del Rancho Arroyo de San Antonio, Sonoma Co., Calif.]. Land Case Map B-711. *Courtesy of The Bancroft Library, UC Berkeley.*

USGS (U.S. Geological Survey). [1914]1916. Santa Rosa Quadrangle, California: 15 minute series (topographic). 1:62,500.

USGS (United States Geological Survey). 2012. *Evaluation of the groundwater resources of the Petaluma Valley*.

Vasey MC, Parker VT, Callaway JC, et al. 2012. Tidal wetland vegetation in the San Francisco Bay-Delta Estuary. *San Francisco Estuary & Watershed Science* 10(2):1-16.

Wagner DL, Gutierrez CI. 2010. Preliminary geologic map of the Napa 30' x 60' quadrangle, California. California Department of Conservation and California Geological Society. 1:100,000.

Wagner DL, Gutierrez CI. 2017. *Preliminary geological map of the Napa and Bodega Bay 30' x 60' quadrangles, California*. California Geological Survey and California Department of Conservation. Sacramento, California.

Watson EB, Byrne R. 2009. Abundance and diversity of tidal marsh plants along the salinity gradient of the San Francisco Estuary: implications for global change ecology. *Plant Ecology* 205:113-128.

Watson EB. 2012. Geomorphology, hydrology, and tidal influence. In *Ecology, Conservation, and Restoration of Tidal Marshes: The San Francisco Estuary*, ed. Arnas Palaima. Berkeley, California: University of California Press.

Wieslander AE. 1930. *Wieslander Vegetation Type Mapping (VTM)*. http://vtm.berkeley.edu/#/data/vegetation

WRCC (Western Regional Climate Center). 1893-2016. *Petaluma AP, California (046826) period of record monthly climate summary*. https://wrcc.dri.edu/cgi-bin/cliMAIN.pl?ca6826

APPENDIX: SPECIES NAMES

Common Name	Scientific Name
Plants	
Alder	*Alnus* spp.
Alkali bulrush	*Bolboschoenus maritimus*
Alkali heath	*Frankenia salina*
Alkali marsh ragwort	*Senecio hydrophilus*
Arroyo willow	*Salix lasiolepis*
Baltic rush	*Juncus balticus*
Beardless wild rye	*Elymus triticoides*
Blackberry	*Rubus ursinus*
Bog yellow cress	*Rorippa palustris*
Buckeye	*Aesculus californica*
Bur reed	*Sparganium eurycarpum* var. *greenei*
California bay laurel	*Umbellularia californica*
California cordgrass	*Spartina foliosa*
California damasonium	*Damasonium californicum*
California rose	*Rosa californica*
Cattail	*Typha* spp.
Coastal button-celery	*Eryngium armatum*
Common blennosperma	*Blennosperma nanum* var. *nanum*
Congested-headed hayfield	*Hemizonia congesta* subsp. *congesta*
Curvepod yellow cress	*Rorippa curvisiliqua*
Douglas' meadowfoam	*Limnanthes douglasii* subsp. *douglasii*
Elderberry	*Sambucus* spp.
False waterpepper	*Persicaria hydropiperoides*
Fleshy jaumea	*Jaumea carnosa*
Gooseberry	*Ribes* spp.
Hairy gumweed	*Grindelia hirsutula*
Harlequin lotus	*Hosackia gracilis*
Hickman's cinquefoil	*Potentilla hickmanii*
Honeysuckle	*Lonicera* spp.
Johnny nip	*Castilleja ambigua* subsp. *ambigua*
Laurel	*Umbellularia californica*
Lobb's aquatic buttercup	*Ranunculus lobbii*

Common Name	*Scientific Name*
Plants, continued from previous page	
Meadow barley	*Hordeum brachyantherum* subsp. *brachyantherum*
Mule fat	*Baccharis salicifolia*
Narrowleaf willow	*Salix exigua* var. *hindsiana*
Oaks	*Quercus* spp.
Oregon ash	*Fraxinus latifolia*
Oregon woolly marbles	*Psilocarphus oregonus*
Pacific woodrush	*Luzula macrantha*
Peppergrass	*Lepidium nitidum*
Petaluma popcornflower	*Allocarya vestita*
Picklweed	*Sarcocornia pacifica*
Pink star-tulip	*Calochortus uniflorus*
Pitkin marsh lily	*Lilium pardalinum* subsp. *pitkinense*
Point Reyes checkerbloom	*Sidalcea calycosa* subsp. *rhizomata*
Red willow	*Salix laevigata*
Salt grass	*Distichlis spicata*
Salt marsh sand spurry	*Spergularia salina*
Sea beet	*Beta vulgaris*
Sea milkwort	*Lysimachia maritima*
Smooth tidy tips	*Layia chrysanthemoides*
Soft bird's beak	*Chloropyron molle* subsp. *molle*
Spike bent grass	*Agrostis exarata*
Stalked popcornflower	*Plagiobothrys stipitatus*
Suisun marsh aster	*Symphyotrichum lentum*
Timothy canary grass	*Phalaris angusta*
Tule	*Schoenoplectus* spp.
Two-fork clover	*Trifolium amoenum*
Water chickweed	*Montia fontana*
Widgeongrass	*Ruppia maritima*
Willows	*Salix* spp.
Yellow rayed goldfields	*Lasthenia glabrata* subsp. *glabrata*
Birds	
American wigeon	*Anas americana*
Black rail	*Laterallus jamaicensis coturniculus*
Bullock's oriole	*Icterus bullockii*
California quail	*Callipepla californica*
Canvasback	*Aythya valisineria*
Hermit thrust	*Catharus guttatus nanus*
Killdeer	*Charadrius vociferus*
Lazuli bunting	*Passerina amoena*
Least sandpiper	*Calidris minutilla*

Common Name	*Scientific Name*
Birds, continued from previous page	
Lincoln's sparrow	*Melospiza lincolnii*
Marsh wren	*Cistothorus palustris*
Northern harrier	*Circus cyaneus*
Northern pintail	*Anas acuta*
Northern Shoveler	*Anas clypeata*
Nuttall's woodpecker	*Picoides nuttallii*
Ridgway's rail	*Rallus obsoletus*
San Pablo song sparrow	*Melospiza melodia samuelis*
Semipalmated plover	*Charadrius semipalmatus*
Short-eared owl	*Asio flammeus*
Song sparrow	*Melospiza melodia gouldii*
Sora	*Porzana carolina*
Spotted sandpiper	*Actitis macularia*
Teals	*Anas* spp.
Tricolored blackbird	*Agelaius tricolor*
Western sandpiper	*Calidris mauri*
Western screech owl	*Megascops kennicottii*
White-tailed kite	*Elanus leucurus*
Willow flycatcher	*Empidonax traillii*
Wilson's snipe	*Gallinago delicata*
Yellow warbler	*Setophaga petechia*
Yellowlegs	*Tringa* spp.
Mammals	
California lowland mink	*Neovison vison aestuarina*
California vole	*Microtus californica*
Northern raccoon	*Procyon lotor*
Ornate shrew	*Sorex ornatus ornatus*
Salt marsh harvest mouse	*Reithrodontomys raviventris halicoetes*
Sonoma shrew	*Sorex vagrans sonomae*
Woodrat	*Neotoma fuscipes*
Amphibians	
California red-legged frog	*Rana draytonii*
Foothill yellow-legged frog	*Rana boylii*
Fish	
Central California Coast steelhead	*Oncorhynchus mykiss irideus*
Riffle sculpin	*Cottus gulosus*
Sacramento sucker	*Catostomus occidentalis*
Starry flounder	*Platichthys stellatus*
Sturgeon	*Acipsenser* spp.
Three-spine stickleback	*Gasterosteus aculeatus*

www.ingramcontent.com/pod-product-compliance
Lightning Source LLC
Chambersburg PA
CBHW041441010526
44118CB00003B/144